Instructor's Manual

to accompany

BASIC STATISTICAL CONCEPTS

Fourth Edition

Albert E. Bartz
Concordia College

Merrill,
an imprint of Prentice Hall
Upper Saddle River, New Jersey   *Columbus, Ohio*

Printed in the United States of America

10 9 8 7 6 5 4 3 2 1

ISBN: 0-13-758988-3

INTRODUCTION

If you would query a random sample of 50 teachers of statistics, you would probably find 50 different philosophies of testing and almost as many types of tests. We would encounter such issues as open versus closed-book tests, objective versus essay items, multiple-choice versus true-false items, and objective items versus work problems. A good case may be made for any of these options, although my personal preference leans towards tests with objective, multiple-choice items.

Our semester course has five competency-graded lab periods where students receive credit upon completing an analysis on data collected in class from several personality measures and college-life demographics. These exercises, plus optional "homework" exercises for each chapter, give the student what I feel is ample practice with work problems and manipulation of formulas. As a result, when constructing a test, I prefer to employ multiple-choice items that assess the students' grasp of particular concepts. While recognizing the importance of being able to work problems, I believe that the multiple-choice items are particularly suited for teasing out the students' ability to make fine discriminations, and give evidence of their grasp of a concept.

Because a first course in statistics seems difficult to many students, I like to test quite frequently. This is definitely not the course where only a midterm exam and a final exam can be used to evaluate a student's performance accurately. Frequent testing and immediate feedback of results are essential in helping the students see how they are doing. Difficulties that an individual student is having are much more likely to be detected when the instructor has frequent samples of each student's performance.

This manual contains a substantial number of items for most chapters to permit frequent testing. Some of the items have been reworked or reworded from the Instructor's Manual, 3rd Edition, while other items are completely new. I thank the many students and colleagues who, through frank and open discussion, have helped me refine many of the items. And I am grateful to Betty Rein for her assistance in composing this edition of the test bank.

Prentice Hall: Instructor Support for Test Item Files

This hard copy test item file is just one part of Prentice Hall's comprehensive testing support service, which also includes:

1.   **Prentice Hall Custom Test:** This powerful computerized testing package is designed to operate on the DOS, WINDOWS, and MACINTOSH platforms. It offers full mouse support, complete question editing capabilities, random test generation, graphics, and printing capabilities.

Prentice Hall Custom Test has a unique two-track design---*Easytest* for the novice computer user, and *Fulltest* for those who wish to write their own questions and create their own graphics.

The built-in algorithmic module enables the instructor to create thousands of questions and answers from a single question template.

In addition to traditional printing capabilities, Prentice Hall Custom Test also offers the On-Line Testing System---the most efficient, time-saving examination aid on the market. With just a few keystrokes, the instructor can administer, correct, record, and return computerized exams over a variety of networks.

Prentice Hall Custom Test is designed to assist educators in the recording and processing of results from student exams and assignments. Much more than a computerized gradebook, it combines a powerful database with analytical capabilities so the instructor can generate a full set of statistics. There is no grading system more complete or easier to use.

The Prentice Hall Custom Test is free. To order a specific Prentice Hall Custom Test title, you may contact you local rep or call our Faculty Support Services Department at 1-800-526-0485. Please identify the main text author and title.

Toll-free **technical support** is offered to all users at **1-800-550-1701.**

2.   For those instructors without access to a computer, we offer the popular **Prentice Hall Telephone Testing Service:** It's simple, fast, and efficient. Simply pick the questions you'd like on your test from this bank, and call the Simon & Schuster Testing Service at 1-800-550-1701; outside the U.S. and Canada, call 612-550-1705.

Identify the main text and test questions you'd like, as well as any special instructions. We will create the test (or multiple versions if you wish) and send you a master copy for duplication within 48 hours. Free to adopters for life of text use.

CONTENTS

CHAPTER 1
SOME THOUGHTS ON MEASUREMENT

1. An admissions office, in describing entering freshmen, would most likely employ which of the following types of statistics?
   a. Inferential
   b. Generalization
   c. Cumulative
   d. Descriptive

2. The use of polls to predict the outcome of an election is an example of:
   a. parametric statistics.
   b. inferential statistics.
   c. nonparametric statistics.
   d. descriptive statistics.

3. The most basic operation involved in any kind of measurement is that of:
   a. adding.
   b. differentiating.
   c. equalizing.
   d. ordering.

4. The scale that allows one to differentiate between classes without ranking them is called a(n):
   a. ordinal scale.
   b. ratio scale.
   c. interval scale.
   d. nominal scale.

5. If I were to classify everybody in this room according to their political affiliation, I would be constructing a(n):
   a. ordinal scale.
   b. nominal scale.
   c. ratio scale.
   d. interval scale.

6. The simplest scale for which one can discriminate differences and show that these differences are of the nature "more than" or "less than" is the:
   a. nominal.
   b. interval.
   c. ratio.
   d. ordinal.

7. The centigrade scale developed for the measurement of heat is an example of:
   a. a nominal scale.
   b. an ordinal scale.
   c. a ratio scale.
   d. an interval scale.

8. The only number scale that possesses an absolute zero point is the:
   a. interval scale.
   b. ratio scale.
   c. ordinal scale.
   d. nominal scale.

9. "One object is four times as much as another object." This can be said only of objects which are measured on:
   a. an ordinal scale.
   b. a ratio scale.
   c. an interval scale.
   d. a nominal scale.

10. $P_1 + P_2 + P_3 + P_4$ expressed in summation form would be:

   a. $4(P_i)$.

   c. $\sum_{i=1}^{4} P_i$

   b. $\sum P_1 + \sum P_2 + \sum P_3 + \sum P_4$.

   d. $a(\sum P_i)$.

11. The "general case" for an observation in the variable X would be
   a. X_1
   b. X_i
   c. X_N
   d. X

12. The size of a group of observations is denoted as
   a. $\sum X$
   b. i
   c. N
   d. X_N

13. If one is interested in making general conclusions about a population from the samples drawn from that population, it would be best to use:
   a. descriptive statistics.
   b. deductive statistics
   c. inferential statistics.
   d. graphical representation of the data.

14. Automobile license plate numbers constitute a(n) _____ scale.
   a. nominal
   b. ordinal
   c. interval
   d. ratio

15. Scales measuring time, height, and weight must have
   a. equal intervals.
   b. an absolute zero.
   c. either a or b.
   d. both a and b.

16. Test scores, in terms of number of items correct, are usually interpreted as representing a(n) _____ scale.
   a. nominal
   b. ordinal
   c. interval
   d. ratio

17. The number "6" may be a glove size, placement in a race, or quantity of apples. Meaningful interpretation is possible only if we use the appropriate
   1. measurement scales.
   2. descriptive statistics.
   c. arithmetic functions.
   d. summation notation.

18. The main difference between descriptive and inferential statistics is that in inferential statistics, we are more interested in
   a. prediction.
   b. consistency.
   c. accuracy.
   d. scaling.

19. We calculate "statistics from statistics by statistics." In this course we will be concerned mostly with statistics as
    a. the results of computations.
    b. raw data.
        e. all of the above.
    c. process or method.
    d. both a and c.

20. A term often used with inferential statistics is
    a. achievement.
    b. incidence.
    c. sampling.
    d. vital.

21. The symbol X_i represents
    a. any score in a column of scores.
    b. the first score in a column.
    c. the last score in a column.
    d. the sum of all of the scores.

22. Data that have been tabulated into categories such as employed/unemployed or single/married/divorced/widowed would constitute a(n) _____ scale.
    a. nominal
    b. ordinal
    c. interval
    d. ratio

23. According to the author, reducing large masses of data to some meaningful values
    a. is the task of statistics.
    b. is the definition of measurement.
    c. differentiates descriptive and inferential statistics.
    d. requires interval or ratio scaling.

24. The symbol $\sum_{i=1}^{5} X_i$ indicates that you are to sum
    a. scores 1 through 5.
    b. scores 1 through 4.
    c. scores 1 and 5.
    d. all the scores.

25. The sum of a constant times a variable is equal to the constant times the sum of the variable. Which expression below illustrates this rule?
    a. $\Sigma(X + Y) = \Sigma X + \Sigma Y$
    b. $\Sigma C^2 = NC^2$
    c. $\Sigma(X - Y)^2 = \Sigma X^2 - 2\Sigma XY + \Sigma Y^2$
    d. $\Sigma(X + Y + C) = \Sigma X + \Sigma Y + NC$

26. According to the summation rules, which expression is the final form of $\Sigma(Y + C)^2$?
    a. $\Sigma Y^2 + \Sigma 2CY + \Sigma C^2$
    b. $\Sigma Y^2 + 2C\Sigma Y + NC^2$
    c. $\Sigma Y^2 + 2NC\Sigma Y + NC^2$
    d. $\Sigma Y^2 + 2\Sigma CY + \Sigma C^2$

27. The expression "$X_3 = 29$" means that
    a. the sum of the first three scores is 29.
    b. three people scored 29.
    c. the third score is a 29.
    d. 29 students took Test #3.

28. Tara finished third in the 100 meter dash. Her third place finish would be part of a(n) _____ scale.
    a. nominal
    b. ordinal
    c. interval
    d. ratio

29. For a scale to be a ratio scale, it must have
    a. equal intervals.
    b. an absolute zero.
    c. either a or b.
    d. both a and b.

30. Interval scales would be present in which o the following?
    a. Fahrenheit temperature
    b. Our scores on this quiz (number correct)
    c. Social security numbers
    d. Two of the above
    e. All of the above

31. Which situation below is probably not an example of descriptive statistics?
    a. Chapter quiz scores for high school sophomores taking history
    b. ACT or SAT exam scores for high school seniors
    c. Mathematics exam scores for eighth graders
    d. Heights and weights for a college hockey team

32. One graphical method that can easily be misleading in interpretation is the
    a. bar graph.
    b. line graph.
    c. pictogram.
    d. pie chart.

33. We calculate "statistics from statistics by statistics." The second term "statistics" refers to
    a. the results of computations.
    b. method or process.
    c. raw data.
    d. theoretical foundations.

34. Whether a given statistic is descriptive or inferential ultimately depends on
    a. its accuracy.
    b. its measurement scale.
    c. the number of cases.
    d. its purpose.

35. In a distribution of 50 scores, the symbol $\sum_{i=1}^{N} X_i$ is
    a. the same quantity as $\sum X$.
    b. the sum of all the scores.
    c. equal to 50.
    d. two of the above.
    e. all of the above.

36. Categorizing your data as male/female or freshman/sophomore/junior/senior would involve a(n) _____ scale.
    a. nominal
    b. ordinal
    c. interval
    d. ratio

4

37. According to the author, reducing large masses of data to some meaningful values
    a. is the task of statistics.
    b. is the definition of measurement.
    c. differentiates descriptive and inferential statistics.
    d. requires interval or ratio scaling.

38. The symbol $\sum_{i=3}^{8} X_i$ indicates that you are to sum
    a. scores 3 through 8.
    b. scores 3 through 7.
    c. scores 3 and 8.
    d. all eight scores.

39. The expression "$X_9 = 71$" means that
    a. the sum of the first nine scores is 71.
    b. nine people scores 71.
    c. the ninth score is a 71.
    d. 71 students took quiz #9.

40. The term statistics refers to
    a. a particular method.
    b. raw data.
    c. the end product.
    d. all of these.

41. Which of the following would not be an example of descriptive statistics in the high school you attended?
    a. all sophomores receive measles vaccine.
    b. all geometry students take a unit exam.
    c. all seniors take a college entrance exam.
    d. all bookkeeping students keep a record of home expenses for a semester.

42. The type of statistics most likely utilized by the Bureau of the Census would be
    a. graphical.
    b. descriptive.
    c. inferential.
    d. categorical.

43. "Assigning numbers to objects or events according to certain prescribed rules" is a formal definition of
    a. measurement.
    b. statistics.
    c. quantification.
    d. differentiation.

44. On an interval scale, numbers that are assigned will
    a. show differences in categories.
    b. show order along a continuum.
    c. indicate the distance between objects.
    d. all of the above.

5

45. In order to discriminate differences and to note that one quantity is "more than" or "less than" another, you need at least a(n) _____ scale.
    a. nominal
    b. ordinal
    c. interval
    d. ratio

46. As a participant in a marathon run, you are assigned number 712 to attach to your shirt. The numbering of the contestants in this marathon would be an example of a(n) _____ scale.
    a. nominal
    b. ordinal
    c. interval
    d. ratio

47. A statement that says that one object is "twice as much" as another implies the use of the _____ scale.
    a. nominal
    b. ordinal
    c. interval
    d. ratio

48. The judges of the "Miss Pre-Teen Queen" contest decide that contestant number 43 is the winner. Their decision would require at least a(n) _____ scale.
    a. nominal
    b. ordinal
    c. interval
    d. ratio

49. The phrase most closely associated with the nominal scale would be
    a. measurement interval.
    b. difference category.
    c. greater or less than.
    d. high-low ratio.

50. The expression "X_N" indicates
    a. a general term for any score.
    b. the score made by the last student in a column of scores.
    c. the sum of the scores of all students.
    d. the score made by any student.

51. A pictogram is often misinterpreted because the reader is influenced by the
    a. distance between the figures.
    b. relative length of the figures.
    c. relative height of the figures.
    d. relative space occupied by the figures.

52. The task of statistics, according to the text, is to
    a. prevent abuses in reporting data.
    b. reduce large masses of data to some meaningful values.
    c. prevent the "garbage in - garbage out" phenomenon.
    d. make predictions from small amounts of data.

53. "Garbage in - garbage out" means that your results can only be as good as your
    a. statistics (raw data).   c. statistics (end product).
    b. statistics (method).   d. all of the above.
                e. none of the above.

54. In order to determine whether a specific statistic is <u>descriptive</u> or <u>inferential</u>, we would probably want to examine its
    a. sample size.   c. precision.
    b. purpose.   d. relevance.

55. Inferential statistics would most likely be used in
    a. taking a poll to predict a senatorial election.
    b. using an arithmetic test in the 5th grade to determine who could pass an algebra course.
    c. finding if college women who belonged to sororities having a lower divorce rate than those who did not.
    d. all of the above use inferential statistics.
    e. none of the above.

56. When we are told that we should not "add apples and oranges," we are being reminded that
    a. we cannot manipulate numbers blindly, without knowing what they represent.
    b. measurement in the physical sciences is more precise than in the social sciences.
    c. misleading conclusions can be drawn from statistics.
    d. all of the above.

57. The Fahrenheit temperature scale for the measurement of heat is an example of the _____ scale.
    a. nominal   c. interval
    b. ordinal   d. ratio

58. An elementary school teacher is asked to rank the children in the class in order of their "creativity." Such a task would require at least a(n) _____ scale.
    a. nominal   c. interval
    b. ordinal   d. ratio

59. On an aptitude test, four students score: Judy = 60, Greg = 57, John = 42, Brad = 39. If the difference in ability between John and Brad is the same as the difference between Judy and Greg, we are dealing with at least a(n) _____ scale.
    a. nominal   c. interval
    b. ordinal   d. ratio

60. Equality of units is an important characteristic of the _____ scale.
    a. nominal   c. interval
    b. ordinal   d. ratio

61. The system used to assign social security numbers would involve a(n) _____ scale.
    a. nominal
    b. ordinal
    c. interval
    d. ratio

62. The TV weather report predicts a high of 40 degrees for tomorrow. We cannot say that this is only half as warm as a temperature of 80 degrees because this scale lacks the properties of a(n) _____ scale.
    a. nominal
    b. ordinal
    c. interval
    d. ratio

63. The expression "$X_5 = 73$" means that
    a. the sum of the first five scores is 73.
    b. five people scored 73.
    c. the fifth score is a 73.
    d. 73 students took Test #5.

64. The expression $\sum\limits_{i=5}^{N} X_i$ indicates
    a. you are to add the first five scores.
    b. the fifth score.
    c. you are to add all five scores.
    d. you are to add all N scores except the first four.

CHAPTER 2
FREQUENCY DISTRIBUTIONS AND GRAPHICAL METHODS

1. Listing every possible score on an examination and the number of students obtaining each score results in a _____ frequency distribution.
   a. simple
   b. grouped
   c. descriptive
   d. interval

2. 43, 67, 92, 24, 89, 84 - - The range of this group of scores is:
   a. 41.
   b. 69.
   c. 68.
   d. none of the above.

3. A rule of thumb often given as to the number of intervals one should have in a grouped frequency distribution is that the number of intervals should be between:
   a. 8-15.
   b. 20-30.
   c. 10-20.
   d. 15-25.

4. The midpoint of the interval 30-39 is:
   a. 34.5
   b. 35
   c. 35.5
   d. 35.9

5. Which of the following is continuous data?
   a. The number of home runs hit by a major league team
   b. The frequency of books bought at a certain store in a particular week
   c. A golf score
   d. The time it takes someone to run the mile

6. Which of the following is likely to yield discrete data?
   a. Number of telephones per home in New York
   b. Weights of members of the wrestling team
   c. High temperatures in Needles, California, each day of 1998
   d. Test scores on the first introductory psychology exam.

7. What are the upper and lower real limits of the interval ranging from 1.05-1.09?
   a. .55-1.59
   b. 1.00-1.14
   c. 1.055-1.085
   d. 1.045-1.095

8. The custom in graphing frequency distributions is to let the vertical axis represent:
   a. scores or measures.
   b. the midpoints of the class intervals.
   c. the upper limits of the class intervals.
   d. frequencies.

9. It is suggested that one puts the _____ on the horizontal axis of a graph.
   a. dependent variable
   b. independent variable
   c. discrete data
   d. continuous data

10. The Y-axis should be approximately:
    a. the same length as the X-axis.
    b. twice as long as the X-axis.
    c. three fourths as long as the X-axis.
    d. one fourth as long as the X-axis.

11. The width of the bar in the histogram extends between:
    a. the lower and upper real limits of each class interval.
    b. the lower and upper apparent limits of each class interval.
    c. the midpoints of two adjacent intervals.
    d. All of these.

12. In constructing a frequency polygon, the frequency for a given interval is plotted at the:
    a. upper limit of the interval.
    b. midpoint of the interval.
    c. across the entire interval.
    d. point arbitrarily determined by the individual.

13. Kurtosis refers to:
    a. symmetry.
    b. frequency.
    c. flatness or peakedness.
    d. height.

14. A special type of distribution where the identity of each score in a distribution is shown is the
    a. cumulative frequency distribution.
    b. stem-and-leaf display.
    c. grouped frequency distribution.
    d. skewed distribution.

15. Which type of distribution(s) show the identity of each individual score?
    a. Cumulative frequency distribution
    b. Stem-and-leaf display
    c. Simple frequency distribution
    d. Two of the above
       e. All of the above

16. In a stem-and-leaf display of 52 vocabulary scores that range from 29 to 63, the "stems" would vary from
    a. 0 to 9
    b. 20 to 60
    c. 3 to 9
    d. 2 to 6

17. One can easily tell the difference between a grouped frequency distribution and a simple frequency distribution by examining
    a. the size of the interval.
    b. the size of N.
    c. the range of the scores.
    d. two of the above.
       e. all of the above.

18. The upper real limit of the 10-19 interval is
    a. 14.5.
    b. 19.0.
    c. 19.5.
    d. 20.0.

19. The size of the interval, i, in the 6.55-6.64 interval is
    a. 0.09.
    b. 0.1.
    c. 1.
    d. 10.

20. The midpoint of the 6.55-6.64 interval is
    a. 6.56
    b. 6.59
    c. 6.595
    d. 6.60

21. An example of continuous data would be a record of
    a. number of dental cavities for children using fluoride mouthwash.
    b. incidence of flu reported at a university health service.
    c. time to complete the race for all finishers in a marathon.
    d. errors made by American League outfielders.

22. In graphing the frequency distribution, you would let the vertical axis represent
    a. scores.
    b. midpoints.
    c. exact upper limits.
    d. frequencies.

23. A graph showing a histogram or frequency polygon should have the vertical axis about _____ of the horizontal axis.
    a. 1/4
    b. 1/3
    c. 1/2
    d. 3/4

24. If we say that a group of measurements is "normally distributed," we mean that measurements are
    a. made on a random sample of a large number of individuals.
    b. made on a group of normal individuals.
    c. made on a group of average individuals.
    d. distributed in a way that closely approximates the mathematical model of a normal curve.

25. If there is a concentration of scores at the low end of the distribution, there is evidence of
    a. positive skewness.
    b. negative skewness.
    c. kurtosis.
    d. normality.

26. In a negatively skewed distribution, the majority of the scores are located
    a. at the high end of the distribution.
    b. at the low end of the distribution.
    c. at the center of the distribution.
    d. cannot say without further information.

27. Positive skewness is shown in a distribution where the majority of scores are
    a. at the high end of the distribution.
    b. at the low end of the distribution.
    c. at the center of the distribution.
    d. at either end of the distribution.

28. A distribution of scores that has a higher and sharper peak than a normal curve is called
    a. platykurtic.
    b. leptokurtic.
    c. mesokurtic.
    d. a J curve.

29. A J curve demonstrates an extreme degree of
    a. measurement error.
    b. kurtosis.
    c. skewness.
    d. variation.

30. A researcher is studying the effect of habitual cocaine use on memory.
    The _____ would be the _____ variable.
    a. amount of material recalled - - dependent
    b. extent of drug use - - dependent
    c. material recalled - - independent
    d. none of the above

31. The lower real limit of the 30-49 interval is
    a. 29.0.
    b. 29.5.
    c. 30.0.
    d. 30.05.

32. The midpoint of the 400-599 interval would be
    a. 499.5.
    b. 500.
    c. 500.5.
    d. none of these.

33. The size of the interval, i, in the 6.9-7.1 interval is
    a. 0.2.
    b. 0.3.
    c. 2.
    d. 3.

34. Which of the following is not an example of continuous data?
    a. Beth is 21 years old.
    b. Mary loses 14 pounds on her diet.
    c. Rita can run to the campus and back in 10 minutes.
    d. Jo has 6 parking tickets.

35. The horizontal axis of a graph should display
    a. the independent variable.
    b. the dependent variable.
    c. continuous data.
    d. discrete data.

36. The terms "histogram" and "frequency polygon" would be most closely identified with the term
    a. bar graph.
    b. functional relationship.
    c. frequency distribution.
    d. normal curve

37. A distribution of family incomes, checking account balances, and savings deposits are typically
    a. platykurtic.
    b. normally distributed.
    c. positively skewed.
    d. negatively skewed.

38. According to the author, the distribution of test grades in most college classes would be
    a. negatively skewed.
    b. positively skewed.
    c. normally distributed.
    d. leptokurtic.

39. Many colleges and universities have one or more sections of "math refresher" courses for students with poor preparation or ability in mathematics. If these students were given a nationally standardized college algebra test, we would expect a _____ distribution.
    a. negatively skewed
    b. positively skewed
    c. normal
    d. bimodal

40. If a distribution of scores is flatter than the normal curve, it is called
    a. a J curve.
    b. mesokurtic.
    c. leptokurtic.
    d. platykurtic.

41. The experiment cited in the text where drivers' behavior at a stop sign was assessed when a police car was either present or absent was an example of
    a. kurtosis.
    b. asymmetry.
    c. normality.
    d. a J curve.

42. In a study to measure the effect of drug dosage on pain perception, the independent variable would be the
    a. amount of pain perceived by each subject.
    b. amount of drug administered.
    c. speed of withdrawal from the painful stimulus.
    d. both a and c.

43. A simple frequency distribution should have an interval size of
    a. one.
    b. two.
    c. three.
    d. depends on the range.

44. The first step in constructing a frequency distribution would be to determine the
    a. interval frequencies.
    b. interval size.
    c. number of intervals.
    d. range of scores.

45. Which statement below illustrates discrete data?
    a. heights of 7th graders
    b. weights of HS wrestlers
    c. divorce rate of college graduates
    d. finish times for '94 Boston marathoners

46. The histogram resembles a _____ while the frequency polygon is a _____.
    a. bar graph; line graph
    b. bar graph; normal curve
    c. box plot; line plot
    d. line graph; bar graph

The next five items are based on this distribution.

Scores	f
70-79	2
60-69	4
50-59	7
40-49	11
30-39	6
20-29	5
10-19	3
0-9	1
	39

47. The largest value that the range of these scores could have is
    a. 9
    b. 39
    c. 70
    d. 79

48. The interval size, i, shown above is
    a. 5
    b. 9
    c. 10
    d. 39

49. The apparent limits of the 60-69 interval are
    a. 59.5-69.5
    b. 60-69
    c. 60-70
    d. 59-69

50. The real limits of the 60-69 interval are
    a. 59.5-69.5
    b. 60-69
    c. 60-70
    d. 59-69

51. The midpoint of the 60-69 interval is
    a. (69-60)/2
    b. 64
    c. 64.5
    d. 65

52. Which interval size are you <u>not</u> likely to see in a grouped frequency distribution?
   a. two
   b. three
   c. four
   d. ten

53. An example of "counting" data, which are treated as continuous data would be
   a. percent of votes cast for the winning candidate
   b. test scores measured as number of items answered correctly
   c. number of HIV positive cases reported for a given state
   d. two of the above
   e. all of the above

54. Most graphs of the frequency distribution take the form of the
   a. functional relationship.
   b. frequency polygon.
   c. histogram.
   d. pie chart.

55. An example of a dependent variable would be the
   a. brightness of a light.
   b. loudness of a sound.
   c. percent of words recalled in a memory test.
   d. two of the above.
   e. none of the above

56. A measure of social conformity would probably show up in a graph as
   a. a normal curve.
   b. extreme platykurtosis.
   c. extreme leptokurtosis.
   d. extreme skewness.

Items 57-61 are based on this distribution.

Scores	f
90-99	1
80-89	5
70-79	8
60-69	10
50-59	6
40-49	6
30-39	4
20-29	2
	42

57. The interval size, i, shown above is
   a. 5
   b. 9
   c. 10
   d. 42

58. The largest value that the range of these scores could have is
    a. 9 c. 70
    b. 42 d. 79

59. The apparent limits of the 30-39 interval are
    a. 29.5-39.5 c. 30-40
    b. 30-39 d. 29-39

60. The real limits of the 30-39 interval are
    a. 29.5-39.5 c. 30-40
    b. 30-39 d. 29-39

61. The midpoint of the 30-39 interval is
    a. (39-30)/2 c. 34.5
    b. 34 d. 35

62. The main difference between a percentile rank and a percentile is that a percentile is a(n)
    a. hierarchy. c. proportion.
    b. estimate. d. score.

63. In the plotting of the points in a cumulative percentage curve, the
    a. lower real limits are used.
    b. the lower apparent limits are used.
    c. the upper real limits are used.
    d. the midpoints of the interval are used.

64. The simplest way to approximate the percentile rank of any score from a cumulative percentage graph is to
    a. extend a line from the unit on the base line, up to the cumulative percentage curve, and then over to the vertical axis.
    b. extend a line from the unit on the vertical axis, over to the cumulative percentage curve, and then down to the base line.
    c. follow the cumulative percentage curve from the base line, up to the approximate point required, and then draw a line down to the base line.
    d. None of these.

CHAPTER 3
CENTRAL TENDENCY

1. The mean may be described as that point in a distribution of scores at which:
   a. 50 percent of the cases score above and 50 percent below.
   b. the algebraic sum of the deviations from it is zero.
   c. the greatest number of scores occur.
   d. two of the above.

2. An accurate definition of the mode is
   a. the most frequently occurring score.
   b. the point in a distribution about which the sum of the deviations is equal to zero.
   c. the point below which 50% of the scores lie.
   d. none of the above.

3. The sum of the algebraic deviations from $\overline{X}$ of a distribution.
   a. is equal to zero only when the distribution is normal.
   b. is equal to zero only when the distribution is symmetrical.
   c. is always equal to zero.
   d. changes from distribution to distribution.

4. The mean can be used with what type of data?
   a. Ordinal
   b. Interval
   c. Nominal
   d. All of the above.

5. The correct method for combining a set of means requires:
   a. summing over the means and dividing by the total N.
   b. summing over the means and dividing by the number of means.
   c. summing over the means which have been weighted by their sample size and dividing by total N.
   d. summing over the means which have been weighted by their sample size and dividing by the number of means.

6. The point that divides a distribution so that 50 percent of the data is found above and 50 percent below the point is the:
   a. mean.
   b. mode.
   c. median.
   d. arithmetic average.

7. To compute the median, it is necessary to have at least _____ data.
   a. nominal
   b. ordinal
   c. interval
   d. ratio

8. Given the following scores, find the median. 5, 12, 18, 19, 26, 31.
   a. 18
   b. 18.5
   c. 19
   d. 22.5

9. A frequency distribution with two pronounced humps or areas of high frequency is referred to as being:
   a. bimodal.
   b. symmetrical.
   c. multi-median.
   d. biased.

10. What would be the shape of a distribution with a mean of 55, median of 52, and mode of 50?
    a. Negatively skewed
    b. Symmetrical
    c. Positively skewed
    d. Normal distribution

11. Usually the most preferred measure of central tendency when a distribution is skewed is the:
    a. mean.
    b. median.
    c. mode.
    d. All of these are equally preferred.

12. Stability refers to:
    a. the variability of a set of scores.
    b. the variability of a statistic computed from different samples of the same population.
    c. the variability of different populations.
    d. the usefulness of a variance.

13. To the extent that a sample value is <u>systematically</u> larger or smaller than the population value, the sample value is considered to be:
    a. efficient.
    b. biased.
    c. random.
    d. consistent.

14. Given the scores on an algebra quiz: 7, 4, 5, 4, 3, 4, 1, 2, 5, 2.
    The median for these 10 scores is
    a. 3.50
    b. 3.64
    c. 3.83
    d. 4.00

15. The mean for the scores in the last item above is
    a. 2.80
    b. 3.70
    c. 4.00
    d. 4.33

16. The term "central tendency" is most closely related to the concept
    a. "distribution."
    b. "average."
    c. "value."
    d. "population."

17. _____ is the same as the arithmetic average.
    a. Mode
    b. Median
    c. Mean
    d. All of these

18. $\Sigma(X - \overline{X} = 0$ indicates that the
    a. mean is the middle score in a distribution.
    b. mean is the exact center of the deviations from the mean.
    c. median is the middle score in a distribution.
    d. mode is the most frequent score in a distribution.

19. The median of a group of 8 scores is
    a. half way between the 5th and 6th scores from the bottom.
    b. half way between the 4th and 5th scores from the bottom.
    c. the sum of the scores divided by 8.
    d. the 5th score from the bottom.

20. The median of a group of 7 scores is
    a. half way between the 3rd and 4th scores from the bottom.
    b. half way between the 4th and 5th scores from the bottom.
    c. the sum of the scores divided by 7.
    d. the 4th score from the bottom.

21. Measures of central tendency for several distributions are shown below.
    Which one shows the greatest degree of <u>positive</u> skewness?
    a. mean = 118, median = 109
    b. mean = 60, median = 61
    c. mean = 110, median = 108
    d. mean = 82, median = 88

22. Among measures of central tendency, the mean is used most often because it is the most
    a. stable.
    b. representative.
    c. meaningful.
    d. resistant to skew.

23. When the amount of skewness in a distribution increases
    a. the mean is affected less than the median.
    b. the median is affected less than the mean.
    c. the mean and median are affected equally, but the mode is unchanged.
    d. all measures of central tendency are affected equally.

24. The main difference in calculating $\overline{X}$ and μ is that
    a. $\overline{X}$ will probably be larger.
    b. μ will probably be larger.
    c. N will probably be different.
    d. Both a and c.
        e. Both b and c.

25. When combining several means together to calculate a combined mean, it is necessary to "weight" each group by
    a. multiplying each mean by its N.
    b. dividing each mean by its N.
    c. adding each mean and its N.
    d. adding all the values of N.

The next three items are based on the following eight quiz scores: 9, 11, 9, 8, 12, 9, 10, 8

26. The mean of the eight scores would be
    a. 5.00.
    b. 9.17.
    c. 9.50.
    d. 10.00.

27. The median of the eight scores would be
    a. 9.00.
    b. 9.17.
    c. 9.50.
    d. 10.00.

28. The mode of the eight scores would be
    a. 3.
    b. 8.
    c. 9.
    d. 12.

29. The term $(X - \overline{X})$ in a distribution of scores is
    a. a quantity that is always zero.
    b. how far a score is from the mean.
    c. the average.
    d. a score in that distribution.

30. The 50th percentile is the same as the
    a. mean.
    b. median.
    c. mode.
    d. two of these.

31. Find the median for these scores: 7, 8, 15, 3, 5, 1.
    a. 3.9
    b. 5.5
    c. 6.0
    d. 6.5

32. The most frequently occurring score in a distribution is called the
    a. mean.
    b. median.
    c. mode.
    d. average.

33. If a constant, C, is added to every score in a distribution, the original mean, $\overline{X}$, would now equal
    a. $C\overline{X}$.
    b. $C\Sigma X/N$.
    c. $\overline{X} + C$.
    d. $\overline{X}$ is unchanged.

34. Means and medians are shown below for several distributions. Which one shows the greatest degree of <u>negative</u> skewness?
    a. mean = 98, median = 89
    b. mean = 40, median = 48
    c. mean = 90, median = 67
    d. mean = 62, median = 65

35. When a distribution is markedly skewed, the measure of central tendency that would be the
    a. mean.
    b. median.
    c. mode.
    d. average.

36. If we would take repeated samples from a population, we would find that the _____ fluctuates the least from sample to sample.
    a. mean
    b. median
    c. mode
    d. either 1 or 2

The next three items are based on the following ten observations: 10, 7, 7, 9, 8, 6, 10, 8. 9, 8

37. The median of the ten observations would be
    a. 8.00.
    b. 8.17.
    c. 8.20.
    d. 8.50.

38. The mean of the ten observations would be
    a. 4.00.
    b. 8.00.
    c. 8.17.
    d. 8.20.

39. The mode of the ten observations would be
    a. 3.
    b. 5.
    c. 8.
    d. 10.

40. The term "average" usually refers to the
    a. mean.
    b. median.
    c. mode.
    d. All of these.

41. If a statistic varies little from sample to sample, we say it is
    a. mesokurtic.
    b. resistant.
    c. stable.
    d. unskewed.

42. The "best single value that describes a distribution" is a definition of
    a. the mean.
    b. the median.
    c. the mode.
    d. central tendency.

43. Fifty percent of the scores in a distribution fall below the
    a. mean.
    b. median.

    c. mode.
    d. Both a and b.

44. If a constant, C, is added to each score in a distribution, the measure of central tendency least affected is the
    a. mean.
    b. median.

    c. mode.
    d. All are equally affected.

45. The highest point in the frequency polygon locates the
    a. mean.
    b. median.

    c. mode.
    d. All of these.

CHAPTER 4
VARIABILITY

1. The "range" of a set of weights (in pounds) of football players means
   a. the distance between the 25th and 75th percentiles.
   b. the difference between the two extreme scores.
   c. the cumulative total.
   d. None of the above.

2. The problem with the range as a measure of variability is that its value is
   a. wholly dependent upon just two scores.
   b. affected by skewness.
   c. that it can be employed only with ordinal data.
   d. All of the above.

3. The sum of the absolute values of deviation scores divided by N is equal to:
   a. the range divided by six.
   b. the average deviation.
   c. the standard deviation.
   d. the variance.

4. The result of squaring all the deviations about the mean is:
   a. the sum of these deviations then becomes zero.
   b. all signs become positive.
   c. all signs are unchanged.
   d. all signs become negative.

5. The standard deviation is appropriate for which of the following type of data?
   a. Nominal   c. Interval
   b. Ordinal   d. All of the above.

6. The symbol s^2 is called the _____.
   a. adjusted variation   c. average squared deviation
   b. z score   d. variance

7. Assume a normal distribution $N = 300$. How many cases would one expect to find between ± 2 standard deviations about the mean?
   a. 150   c. 285
   b. 204   d. 297

8. A z score is expressed in:
   a. whole score units.   c. standard deviation units.
   b. ordinal scale units.   d. None of the above.

9. When we know that $\overline{X}$ of a distribution is 25 and s is 4, a score of 35 in the distribution will have a z value of:
   a. 2.75.
   b. 3.00.

   c. 2.50.
   d. 1.25 .

10. If little Johnny weighs 300 pounds and stands 7 feet tall, which of the following is true? (Assume average weight is 150 pounds, average height 5'6", s for weight is 50 pounds and s for height is 6".)
    a. Johnny is taller than he is heavy.
    b. Johnny is heavier than he is tall.
    c. Johnny is as tall as he is heavy.
    d. Cannot say from information given.

11. If the mean of a distribution is 20, and the z for an X score of 26 is 3, what is the standard deviation?
    a. 2
    b. 3

    c. -2
    d. Cannot tell from data given.

12. If a student achieved a score of 10 on a test with a mean of 15 and a standard deviation of 5, and a score of 50 on another test with a mean of 60 and a standard deviation of 10, we could say that:
    a. she achieved better on the first test.
    b. she achieved better on the second test.
    c. she achieved equally well on both tests.
    d. None of the above, as more information is needed.

13. Shown below are the deviations from the mean for seven test scores. What is the standard deviation for the test scores?
    (5, 3, 2, 0, 0, -3, -7)
    a. 0
    b. 4

    c. 12
    d. 16

14. The point of the story of the statistician with feet in the oven and head in the refrigerator was that
    a. a distribution is not completely described by a measure of central tendency.
    b. a measure of central tendency is less stable than a measure of variability.
    c. the mean is not always the best measure of central tendency.
    d. statisticians are notorious for wasting energy.

15. The range is not one of the better measures of variability since it
    a. does not fit the theoretical framework of statistics.
    b. tells us nothing about the size of the distribution.
    c. is insensitive to the size of the mean.
    d. depends on only two scores.

16. Given five scores of 7, 10, 8, 2 and 3, the absolute value of the deviation from the mean for the score of 3 would be
    a. 6.   c. 3.
    b. 0.   d. -3.

17. The symbol $\Sigma (X - \overline{X})^2$ represents which operation?
    a. The deviations from the mean are summed and then squared.
    b. The deviations from the mean are squared and then summed.
    c. The scores are squared and then summed.
    d. The scores are summed and the sum is squared.

18. Two distributions could have the same _____, but different _____.
    a. range; standard deviations
    b. variance; standard deviations
    c. standard deviation; variances
    d. All of the above are possible.

19. Which expression would describe the middle 68% of the distribution in a normal curve?
    a. $\overline{X} \pm s$
    b. $\mu \pm s$
    c. $\mu \pm \sigma$
    d. $\overline{X} \pm \sigma$

20. T, z, and CEEB are all examples of standard scores which means that they
    a. are based on a standard deviation.
    b. are derived from percentile norms.
    c. are based on a normal distribution.
    d. all have a mean of zero.

21. If a test distribution had a mean of 500 and a standard deviation of 100, an individual with a z score of .80 would have a test score of
    a. 420   c. 508
    b. 580   d. 680

22. In calculating a standard deviation for a distribution of scores you find s = -7.83.
    This would indicate
    a. little variation between scores.
    b. large variation.
    c. moderate variation.
    d. a computational error.
    e. Cannot tell without additional information.

23. Given the following information on three nationally standardized tests.
    Test A: $\mu = 80$, $\sigma = 10$; Test B: $\mu = 60$, $\sigma = 5$; Test C: $\mu = 110$, $\sigma = 20$.
    Teri scores 92 on Test A, 67 on Test B and 135 on Test C.
    Which order of the three tests describes her performance from best to worst?
    a. ABC   c. BAC
    b. BCA   d. ACB
                    e. CBA

24. If a constant, C, is added to every score in a distribution whose standard deviation is s, the
    standard deviation of the new distribution would be
    a. C times s   c. C plus s
    b. s plus $\sqrt{C}$ d. unchanged

25. Variability is defined as the
    a. best single value that describes group performance.
    b. variation of the values of the mean, median and mode.
    c. fluctuation of scores about a measure of central tendency.
    d. both a and b.
    e. both a and c.

26. We treat the range as an inferior measure of variability since it
    a. can only be used with ratio data.
    b. is affected by positive or negative skewness.
    c. is dependent on two extreme scores.
    d. all of the above.

27. In the calculation of the average deviation, the absolute values of the deviations are used.
    This is due to the fact that
    a. $\Sigma X = 0$.
    b. $\Sigma X^2 = 0$
    c. $\Sigma (X - \overline{X}) = 0$.
    d. $\Sigma (X - \overline{X})^2 = 0$.

28. When the deviations about the mean are squared
    a. all signs become positive
    b. the sum of the squared deviations is zero.
    c. the signs are unchanged.
    d. Two of the above.
    e. None of the above.

29. The standard deviation requires at least a(n) _____ scale for meaningful interpretation
    a. nominal   c. ordinal
    b. interval   d. ratio

30. In calculating the standard deviation from scores of 6, 7, 9, 6, 5 and 3, we would find s to be equal to
    a. 2.   c. 4.
    b. $\sqrt{20}$.   d. 20.

31. The standard deviation, when applied to the normal curve, would be defined as the distance on the horizontal axis between the
    a. mean and the steepest part of the curve.
    b. steepest parts of the curve on the left and right.
    c. end of the curve and the mean.
    d. left and right ends of the curve.

32. Which measures of variability are mathematically related?
    a. average deviation and variance
    b. average deviation and range
    c. range and standard deviation
    d. standard deviation and variance
    e. average deviation and standard deviation

33. You are asked to calculate the standard deviation for scores of 5, 2, 5, 4, 6, and 8 and come up with s = 2. The <u>variance</u> of this group of scores is
    a. 2   c. $\sqrt{2}$
    b. 4   d. 7

34. $\Sigma(X - \overline{X})^2$ is to $\Sigma(X - \mu)^2$
    a. standard deviation is to variance.
    b. average deviation is to standard deviation.
    c. sample is to population.
    d. deviation formula is to computational formula.

27

35. In a normal curve, approximately 95% of the distribution falls between ±2 _____ .
    a. s
    b. s^2
    c. σ
    d. X

36. A z score of zero would tell us that the score is at the _____ of the distribution.
    a. mean
    b. very top
    c. very bottom
    d. None of the above, since z cannot be zero.

37. If a test distribution has a mean of 50 and a standard deviation of 10, an obtained score of 35 would give a z score of
    a. -.15
    b. 15.
    c. 1.5
    d. None of these.

38. If a test distribution has a mean of 100 and a standard deviation of 20, an individual with a z = -1.30 would have a test score of
    a. 70
    b. 126
    c. 74
    d. 130

39. A weakness of the range as a measure of variability is that
    a. it cannot be used with decimal values.
    b. it does not have a theoretical foundation in statistics.
    c. one extreme value can markedly alter its value.
    d. Two of the above
    e. All of the above

40. Of the terms used to describe variability, which one is most precise?
    a. dispersion
    b. fluctuation
    c. variation
    d. variance

41. The symbol σ represents the
    a. standard deviation of a sample.
    b. standard deviation of a population.
    c. variance of a sample.
    d. variance of a population.

42. A statistics quiz has a mean of 10 and a standard deviation of 2. The z score for a quiz score of 9 would be
    a. 0.5
    b. -0.5
    c. -0.9
    d. -2.0

43. In the quiz above a z score of 2.0 would indicate a quiz score of
   a. 8
   b. 12
   c. 14
   d. cannot say

44. The term <u>variability</u> is related to the degree of
   a. scores varying from the mean.
   b. scores varying from each other.
   c. the difference between the mean and the median.
   d. Two of the above
   e. All of the above

45. Of the terms, "fluctuation, variation, variability, variance, scattering, and dispersion," which term is different from the rest?
   a. dispersion
   b. scattering
   c. variation
   d. variance

46. The symbol ΣX^2 is the same as
   a. $(\Sigma X)^2$
   b. $\Sigma(X - \overline{X})^2$
   c. $[\Sigma X^2 - (\Sigma X)^2]/N$
   d. none of these

47. The symbol σ^2 represents the
   a. standard deviation of a sample.
   b. variance of a sample.
   c. standard deviation of a population.
   d. variance of a population.

48. A math quiz has a mean of 10 and a standard deviation of 2. The z score for a quiz score of 11 would be
   a. 0.1
   b. 0.5
   c. 1.1
   d. 2.0

49. In the quiz above, a z score of -2.0 would indicate a quiz score of
   a. 6
   b. 7
   c. 8
   d. 9

50. If IQ scores are normally distributed in the population, with $\mu = 100$, and $\sigma = 16$, we would expect the middle 95 percent of the scores to be between what two IQ scores?
   _____ and _____

51. In the above distribution what z score corresponds to an IQ of 88? _____

52. In the IQ distribution what score corresponds to a z of 1.75? _____

53. If scores on Mathematics Subtest of the ACT are normally distributed in the population, with $\mu = 18$, and $\sigma = 4$, we would expect the middle 99 percent of the scores to be between what two ACT scores? _____ and _____

54. In the above distribution what z score corresponds to an ACT score of 15? _____

55. In the ACT distribution what score corresponds to a z of 1.25? _____

CHAPTER 5
THE NORMAL CURVE

1. The percentile rank of a z score of 1.0 would be approximately
    a. 16   c. 84
    b. 68   d. 95

2. The unit normal curve has
    a. a mean of zero.
    b. an area equal to one square unit.
    c. a standard deviation of 1.0.
    d. two of the above.
    e. all of the above.

3. To find the z scores marking off the middle 95 percent of the normal curve distribution, we would
    a. find the z scores for .5000 in Table A.
    b. find the z scores for .0500 in Table A.
    c. find the z scores for .9500 in Table A.
    d. None of the above.

4. The odds against rolling a two with a single die are
    a. 5 to 1   c. 1 in 6
    b. 6 to 1   d. 1 to 5

5. To get the area of the curve between z scores of .75 and 1.5, one should:
    a. add the area between the mean and a z of 1.5 to the area between the mean and .75.
    b. subtract the area between the mean and a z of .75 from the area between the mean and 1.5.
    c. subtract z scores.
    d. add z scores.

6. To find the area between a z score of -1.0 and a z score of 1.5, one would:
    a. add the z scores.
    b. subtract the z scores.
    c. add the area from the mean to a z score of -1.00 to the area from the mean to a z score of 1.5.
    d. subtract the area from the mean to a z score of -1.00.

7. The area under the normal curve between a z of +1.0 and +2.0 is:
   a. less than the area between the mean and a z of +1.0.
   b. the same as the area between the mean and a z of +1.0.
   c. greater than the area between the mean and a z of +1.0.
   d. None of the above.

8. The statement, "A 70-year-old American male has p = .29 of living to age 80" is an example of:
   a. empirical probability.
   b. subjective probability.
   c. classical probability.
   d. All of the above.

9. The definition of probability is:
   a. a ratio of the total number of ways all events under consideration can occur divided by the number of ways a specified event can occur.
   b. the number of ways a specified event can occur multiplied by the ways it can't occur.
   c. a ratio of the number of ways a specified event can happen divided by the total number of ways all events under consideration can occur.
   d. a ratio of the number of ways an event can occur divided by the number of ways the event can't occur.

10. Probability is a:
    a. proportion.
    b. percentage.
    c. score.
    d. z score.

11. If we roll a die, the probability of obtaining either a 4 or 5 is:
    a. 1/36.
    b. 1/18.
    c. 1/6.
    d. 1/3.

12. As the probability of a z score increases, the deviation from the mean
    a. increases.
    b. decreases.
    c. remains the same.
    d. depends upon the type of event.

13. A nationally standardized mathematics exam has a mean of 100 and a standard deviation of 20. Approximately what percent of the population would have scores below 80?
    a. 5
    b. 16
    c. 34
    d. 68

14. A nationally standardized English vocabulary exam has a mean of 500 and a standard deviation of 100. If a student scores 328 on the exam, what would the z score be?
    a. 1.28
    b. 1.7
    c. -1.28
    d. -1.72

15. Probabilities concerning a dice throw, a coin toss, or a card draw are called _____ probabilities.
    a. "long run"   c. empirical
    b. certainty   d. classical

16. When we relate the mathematical normal curve to data in the social sciences, we would say that the curve
    a. is an unbiased estimate.   c. describes natural events.
    b. is an ideal.   d. is a model.

17. The first use of the normal curve on actual data involved measurements of
    a. IQ.   c. memory span.
    b. reaction time.   d. soldiers' heights.

18. To determine the area under the normal curve <u>above</u> a positive z score, you would
    a. find the value directly from Table A.
    b. find the area for the given z, and subtract from .5000.
    c. find the area for the given z, and add to .5000.
    d. find the area for the given z, and subtract from 1.000.

19. To calculate the area between a z score of 1.2 and a z score of -.90, we would
    a. add the area from the mean to a z of -.90 to the area from the mean to a z of 1.2.
    b. add the z scores and find the area from the mean to 2.10.
    c. subtract the z scores and find the area from the mean to .30.
    d. subtract the area from the mean to a z of -.90 from 1.000.

20. To determine the area of the curve between z scores of 1.3 and .85, we would
    a. add the area between the mean and a z of 1.3 to the area between the mean and .85.
    b. subtract the area between the mean and a z of .85 from the area between the mean and 1.3.
    c. subtract the z scores and find the area from the mean to .45.
    d. add the z scores and find the area from the mean to 2.15.

21. To find the <u>percentile rank</u> of a given positive z score, you would
    a. look up the value directly from Table A.
    b. find the area for the given z and add it to .5000.
    c. find the area for the given z and subtract it from .5000.
    d. percentile rank cannot be determined unless further information is given.

22. To find the area under the normal curve between two negative z scores, we would
    a. find the value given in the normal curve table.
    b. subtract the smaller from the larger z, and find the area in the table.
    c. find the areas for the respective z scores in the table and add them.
    d. find the areas for the respective z scores in the table, and subtract the smaller area from the larger area.

23. A nationally standardized mathematics exam has a mean of 100 and a standard deviation of 20. Approximately what percent of the population would have scores below 120?
    a. 64
    b. 74
    c. 84
    d. 94

24. The "odds against" drawing an ace from a deck of cards would be
    a. 4 in 52
    b. 4 to 48
    c. 52 to 4
    d. 48 to 4

25. A man is chosen at random from the population. The probability that his weight is between $\mu \pm \sigma$ is
    a. .05.
    b. .68.
    c. .34.
    d. .95.

26. As we plan our picnic, we hear on the weather news that there is a "40 percent chance of rain." This would be an example of
    a. classical probability.
    b. theoretical probability
    c. empirical probability.
    d. average odds.

27. Which statement is true regarding the normal curve?
    a. There are many normal curves, depending on μ and σ.
    b. There are many normal curves, depending on $\overline{X}$ and s.
    c. There are many normal curves, depending on the range and s.
    d. There is one normal curve, bell-shaped and symmetrical about $\overline{X}$.

28. The <u>unit normal curve</u> is so named because
    a. it has a mean equal to one.
    b. its area is equal to one square unit.
    c. it has a standard deviation of one.
    d. Two of the above.
    e. All of the above.

29. The normal curve was first developed by a mathematician to answer questions about
    a. gambling.
    b. the military.
    c. physics.
    d. mental abilities.

30. In the normal curve table, a z value of .75 has a corresponding entry of .2734. This means that 27.34% of the area
    a. is between z values of -.75 and +.75.
    b. is between the mean and a z of .75.
    c. falls above a z of .75.
    d. falls below a z of .75.

31. To calculate the z scores that mark off the middle 30 percent of the area under the normal curve, you would
    a. find the values in the normal curve table.
    b. subtract 30 from 50 percent and look up the values for the remainder in the table.
    c. divide 30 by 2 to obtain 15 percent, subtract from 50 percent and look up the values for the remainder in the table.
    d. divide 30 by 2 to obtain 15 percent, and look up the values in the table.

32. To calculate the area under the normal curve below a positive z score, you would find the
    a. value given in the normal curve table.
    b. area for the z score and subtract from .5000.
    c. area for the z score and add it to .5000.
    d. area for the z score and subtract it from 1.000.

33. To calculate the <u>percentile rank</u> of a negative z score, we would
    a. find the value given in the normal curve table.
    b. find the area for the z score and subtract it from .5000.
    c. find the area for the z score and add it to .5000.
    d. find the area for the z score and subtract it from 1.000.

34. To determine the area between two positive z scores, you would
    a. find the values for the respective z scores in Table A, and subtract the smaller area from the larger area.
    b. locate the values for the respective z scores in Table A, and add them.
    c. look up the value directly from Table A.
    d. subtract the smaller from the larger z score, and find the value in Table A.

35. A college entrance exam has a mean of 100 and a standard deviation of 20. If it is necessary to have a score of at least 80 before you can be admitted to a certain college, approximately what percent of the population is <u>not</u> eligible?
    a. 5
    b. 34
    c. 16
    d. 68

36. A nationally standardized English vocabulary exam has a mean of 500 and a standard deviation of 100. A z score of -1.23 would be a test score of
    a. 377
    b. 498.77
    c. 487.7
    d. 512.3
    e. 623

37. A nationally standardized musical aptitude test has a mean of 100 and a standard deviation of 30. What percent of the population would score above 160?
    a. 1
    b. 5
    c. 2.5
    d. 99

38. A statistician must use previous observations in order to determine
    a. the "long run."
    b. classical probability.
    c. certainty.
    d. empirical probability.

39. The "odds for" drawing a queen of diamonds from a deck of cards would be
    a. 4 in 52.
    b. 4 to 48.
    c. 1 in 52.
    d. 1 to 51.

40. A woman is chosen at random from the population. The probability that her height is outside the region $\mu \pm 1.96\sigma$ is
    a. .01
    b. .50
    c. .05
    d. .95

41. "Predicting" and "visualizing" is possible when we use the normal curve
    a. as an ideal.
    b. to describe 30 quiz scores.
    c. as a model.
    d. All of these.

42. Determining the probability of drawing an ace from a deck of cards would involve _____ probability.
    a. classical
    b. empirical
    c. experimental
    d. mathematical

43. In an appliance store's sales promotion, customers may choose one of 300 numbered balls in a basket, where 6 are especially numbered to give customers a 40% discount on their purchase. If you pick a ball at random, which statement describes your chances of getting the discount?
    a. p = .02; odds for, 2:98
    b. p = .40; odds against, 60:40
    c. p = .02; odds against 300:6
    d. p = .06; odds for, 6:94

44. Using the normal curve as a model is important for
    a. predicting future data.
    b. visualizing the nature of our data.
    c. selecting our data.
    d. two of the above.
    e. all of the above.

45. The key phrase in calculating empirical probabilities is
    a. law of large numbers
    b. normal distribution
    c. number theory
    d. previous observations

46. In a furniture store's sales promotion, customers may choose one of 200 numbered balls in a basket, where 2 are especially numbered to give customers a 20% discount on their purchase. If you pick a ball at random, which statement describes your chances of getting the discount?
    a. p = .01; odds against, 200:2
    b. p = .20; odds for, 20:80
    c. p = .02, odds for, 2:98
    d. p = .01, odds against, 99:1

47. Applying the normal curve to actual data was probably first done by Quetelet with
    a. height and chest measurements.
    b. intelligence test scores.
    c. leading causes of death.
    d. racial and gender differences.

48. The six U.S. coins (penny, nickel, dime, quarter, half dollar, and dollar) are in a jar, and one is drawn at random. What is the probability that this coin is a penny?
    a. 1/100
    b. 5/100
    c. 6/100
    d. 1/6
    e. 1/5

49. In the coin problem above, what is the probability that the coin is either a penny or a dollar?
    a. 2/100
    b. 10/100
    c. 12/100
    d. 2/6
    e. 2/5

50. In a card-drawing exercise, a card is drawn and returned to the deck, and a second card is drawn. What is the probability that both cards are kings?
    a. 1/4 x 1/4
    b. 1/4 + 1/4
    c. 4/52 x 4/52
    d. 4/52 + 4/52

51. In the card-drawing exercise above, what is the probability that both cards are spades?
    a. 1/4 x 1/4
    b. 1/4 + 1/4
    c. 4/52 x 4/52
    d. 4/52 + 4/52

The California Mathematics Aptitude Test (CMAT) has a mean of 500 and a standard deviation of 100. Assume that this variable of math ability is normally distributed in the population of high school seniors. (You will need to use Table A to answer these questions.)

52. What percent of the seniors would be expected to score 350 or less?

53. What percent of the seniors would be expected to score between 365 and 635?

54. Joe is disappointed to learn that he is in the bottom 20% of seniors in math ability. How low would his CMAT score have to be for him to make that statement?

55. The middle 95% of the seniors should score between what two values?

56. What is the probability of a senior, chosen at random, scoring 400 or higher?

57. Students entering college who scored at the 90th percentile or above are given advanced placement in college math requirements. What CMAT score is necessary to be in that grouping?

The Stanford Reading Exam (SRE) has a mean of 100 and a standard deviation of 20. Assume that this variable of reading ability is normally distributed in the elementary school population. (You will need to use Table A to answer these questions.)

58. What percent of the children would be expected to score 135 or higher?

59. What percent of the children would be expected to score between 75 and 125?

60. Kristin's mother says her daughter is in the top 10 percent of children in reading ability. How high would her score have to be for her mother to make this claim?

61. The middle 95% of the children should score between what two values?

62. What is the probability of a child, chosen at random, scoring 110 or less?

63. One of the teacher aids works with children who score less than the 20th percentile. What SRE score would that be?

CHAPTER 6
SAMPLING THEORY FOR HYPOTHESIS TESTING

1. We may expect that the mean of a random sample will <u>not</u> be arithmetically identical with the mean of a population because
   a. of sampling error.
   b. population data are ordinal.
   c. different units of variability are used.
   d. None of these.

2. If we draw all possible samples of the same size from a population, then the mean of all sample means will be
   a. greater than the population mean.
   b. less than the population mean.
   c. equal to the population mean.
   d. an unbiased estimate of the population mean.

3. The major difference between the standard error and standard deviation of a sampling distribution of means is
   a. standard deviation is a measure of variability and standard error isn't.
   b. normal curve cannot be applied to standard error measures.
   c. there is no difference between the two statistics.
   d. standard error measures the extent of error associated with a measure and standard deviation doesn't.

4. The standard error of the mean is an estimate of
   a. the standard deviations for a number of successive distributions from the same population.
   b. the standard deviation of a distribution of randomly drawn sample $\overline{X}$ s.
   c. the range of the means of a distribution of randomly drawn samples from the same population.
   d. the mean of the population from which the samples were drawn.

5. Which of the following will decrease the standard error of the mean?
   a. Increasing the standard deviation of the sample
   b. Increasing N
   c. Increasing the mean
   d. Decreasing the mean

6. In the 95 percent confidence interval
   a. there's a 5 percent chance that the population mean is within the interval.
   b. there's a 95 percent chance that the population mean is within the interval.
   c. 95 percent of such intervals will include the population mean.
   d. 5 percent of the sample means are within the interval.

7. The width of the confidence interval is dependent upon
   a. the standard deviation of the population.
   b. the sample size.
   c. the degree of confidence required by the research worker.
   d. all of the above.

8. If a hypothesized population mean is 60, a sample mean is 62 and $\sigma_{\bar{x}}$ is 10, then the z score for the sample mean is
   a. -.2
   b. 2
   c. .2
   d. Cannot be calculated

9. A developmental psychologist believes that the average math aptitude score for 8th graders is 100. A test is given to a group of 8th graders and the mean is 105 and $\sigma_{\bar{x}} = 10$. What is the 95% CI for the population mean?
   a. 80.4 to 119.6
   b. 95 to 115
   c. 90 to 110
   d. 85.4 to 124.6

10. The confidence interval approach for estimating the population mean assumes that the center of the confidence interval is the
    a. sample mean.
    b. standard error of the mean.
    c. mean of the sampling distribution.
    d. population mean.

11. Sampling error is demonstrated in which of the following examples?
    a. Your watch is always 5 minutes fast.
    b. A friend gets four "A's" in her courses last semester.
    c. Your uncle draws 5 consecutive spades from a well-shuffled deck of cards.
    d. You get sick from tasting different brands of eggnog in a supermarket display.

12. Which of the following illustrates the central limit theorem?
    a. The sample mean approaches the value of the population mean.
    b. Sampling distribution will be normal regardless of the shape of the population.
    c. Sampling error decreases with increasing sample size.
    d. The sample variance is an unbiased estimate of the population variance.

13. The standard error of the mean is
    a. an estimate of the standard deviation of the sampling distribution.
    b. an unbiased estimate of the population mean.
    c. equal to the standard deviation of the sample.
    d. equal to the standard deviation of the population.

14. In actual practice, if we wanted a narrower confidence interval, we would probably
    a. decrease the standard deviation.
    b. increase the standard deviation.
    c. use the 99% instead of the 95% CI.
    d. increase the sample size.

15. A researcher believes that the average vocabulary score for 5th graders is 50. A test is given to a sample of 5th graders, and the mean is 55 and $\sigma_{\bar{x}} = 10$. What is the 95% CI for the population mean?
    a. 30.4 to 69.6
    b. 40 to 60
    c. 35.4 to 74.6
    d. 45 to 65

16. With a 95% confidence interval
    a. 5 percent of the sample means are in the interval.
    b. 95 percent of such intervals will include the population mean.
    c. the probability is .05 that the population mean is in the interval.
    d. the probability is .95 that the population mean is in the interval.

17. In sampling from a large population, the _____ method is very convenient.
    a. correlated
    b. counting off
    c. random number table
    d. parametric

18. A "miniature population" would occur in a _____ sample.
    a. biased
    b. representative
    c. random
    d. two of the above
             e. all of the above

19. A biased sample is one that
    a. is too small.
    b. has a systematic error.
    c. is not random.
    d. lacks independence.

20. Statistic is to parameter as sample is to
    a. population.
    b. random.
    c. unbiased.
    d. representative.

21. The distribution of an infinite number of sample means from a population
    a. is normally distributed.
    b. is called a sampling distribution of means.
    c. has a mean, μ.
    d. Two of the above.
    e. All of the above.

22. In the formula for the standard error of the mean, $\dfrac{\sigma}{\sqrt{N}}$

the N stands for the number of
a. means in the sampling distribution.
b. observations in the population.
c. observations in the sample.
d. means in the population.

23. The widest confidence interval would be given by the _____ CI.
a. 68%
b. 95%
c. 90%
d. 99%

24. A statistician has hypothesized H_o: $\mu = 90$. He then takes a sample and calculates a mean of 75. If $\sigma_{\bar{x}} = 5$, what would the statistician conclude?
a. Reject hypothesis that population mean is 90.
b. Do not reject hypothesis that population mean is 90.
c. Accept hypothesis that population mean is 75.
d. Two of the above are correct.

25. The probability level we would use to say that the difference between a hypothesized population mean and our sample mean is <u>not</u> due to sampling error is the
a. .05
b. .01
c. .001
d. We can never deny the possibility of a sampling error.

26. "Even though the population may be non-normal, the sampling distribution will be normal." This statement is most closely associated with
a. biased estimates.
b. random sampling.
c. unbiased estimates.
d. the central limit theorem.

27. Which of the following statements is false? A population . . . .
a. is a group of elements alike on one or more characteristics.
b. is defined by the researcher.
c. is always very large.
d. consists of all elements satisfying the research criteria.

28. Examples of parameters would be
a. $\overline{X}$ and s
b. $\overline{X}$ and μ
c. s and σ
d. μ and σ

29. The symbol, $\sigma_{\bar{x}}$, represents the
    a. standard deviation of the sample.
    c. mean of the sample.
    b. standard deviation of the population.
    d. mean of the population.
                    e. None of the above.

30. Which of the following equalities is not true?
    a. $\mu_{\bar{x}} = \mu$
    c. $\sigma_{\bar{x}} = \sigma/\sqrt{N}$
    b. $\sigma = \sigma_{\bar{x}}$
    d. All of the above are true.

31. The widely used Clerical Aptitude Test (CAT) has a standard deviation of 20. A technical school announces that its beginning data processing class of 100 students had a mean of 80 on the CAT. Calculate the 95% CI.
    a. 60-100
    c. 76.08-83.92
    b. 77.42-82.58
    d. 78.04-81.96

32. The superintendent believes that the true population mean for the CAT is 75. Based on your calculations, you would
    a. reject H_0 since 75 is inside the interval.
    b. reject H_0 since 75 is outside the interval.
    c. not reject H_0 since 75 is outside the interval.
    d. not reject H_0 since 75 is inside the interval.

33. Which of the following would not be considered a population?
    a. Fourth-graders in private elementary schools in a given state.
    b. College graduates in the U.S. for a given year.
    c. Data entry operators for a large metropolitan school district.
    d. All of the above could be populations.

34. A friend is attempting to calculate the probability of drawing an ace of spades from a deck of cards. She fails on her first three draws and reports that the probabilities for these three draws were 1/52, 1/51, and 1/50. You know that she
    a. is not using random sampling.
    c. has biased samples.
    b. is sampling without replacement.
    d. is using incidental samples.

35. Which of the following statements about the sampling distribution of means is(are) not true?
    a. The mean is equal to μ.
    b. The standard deviation is σ.
    c. The means are normally distributed.
    d. Two of the above.
    e. All of the above.

36. The widely used College Entrance Exam (CEE) has a standard deviation of 20. A small private college announces that its freshman class of 100 had a mean of 105 on the CEE. Calculate the 95% CI and indicate your choice below.
    a. 85-125
    b. 101.08-108.92
    c. 103.04-106.96
    d. 96.08-103.92

37. The college president believes that the true population mean for the CEE is 100. Based on your calculations, you would
    a. reject H_0 since 100 is inside the interval.
    b. not reject H_0 since 100 is inside the interval.
    c. reject H_0 since 100 is outside the interval.
    d. not reject H_0 since 100 is outside the interval.

38. In the one-sample case, the null hypothesis, H_0, states that
    a. the population mean is zero.
    b. the mean of the sampling distribution is zero.
    c. the difference between the population mean and the mean of the sampling distribution is zero.
    d. None of the above.

39. The null hypothesis, in the one sample case, states a hypothetical value for
    a. the population mean.
    b. the mean of the sampling distribution.
    c. the sample mean.
    d. the difference between the population mean and the sample mean.

A sample of 49 professional women has a mean height of 5'7". The Metropolitan Life Tables of heights and weights show a standard deviation of 2.1" for this population of heights.

40. Calculate the standard error of the mean for this data.

41. Calculate the 95% CI for this sample.

42. A friend believes that the average height should be 5'5". According to your CI is this likely? Why or why not?

A sample of 25 sophomore women on a college campus has a mean height of 5'7". The Metropolitan Life Tables of heights and weights show a standard deviation of 3.5" for the population of heights of college women.

43. Calculate the standard error of the mean for this data.

44. Calculate the 95% CI for this sample.

45. A friend believes that the average height of college women is 5'5". According to your CI is this likely? Why or why not?

CHAPTER 7
CORRELATION

1. A positive correlation indicates that as:
   a. one variable increases, the other one also increases.
   b. one variable decreases, the other one also decreases.
   c. one variable increases, the other one decreases.
   d. both (a) and (b).

2. If the family income is the same regardless of family size, we speak of:
   a. positive correlation.
   b. negative correlation.
   c. zero correlation.
   d. Correlation does not apply to this situation.

3. A negative sum of the products of z scores indicates that:
   a. low X scores tend to be associated with high Y scores.
   b. low Y scores tend to be associated with low X scores.
   c. high X scores tend to be associated with high Y scores.
   d. any X scores tend to be associated with either low or high Y scores.

4. Which indicates the greatest degree of relationship?
   a. +.57   c. +.10
   b. -.67   d. -.05

5. A fat ellipse, running from the upper left-hand corner of a graph to the lower right hand corner, indicates:
   a. a high positive correlation.
   b. a low positive correlation.
   c. a high negative correlation.
   d. a low negative correlation.

6. Which of the following relationships is indicated if the tally marks in a scatter diagram form a thin ellipse running from the lower left corner to upper right corner of the diagram?
   a. High positive   c. High negative
   b. Low positive   d. Low negative

7. Significant values of r depend upon:
   a. the direction of the r, whether positive or negative.
   b. the different measurement scales used.
   c. the degrees of freedom.
   d. whether the r is computed from whole scores or from z scores.

8. The Pearson r is appropriate only when the data is at least in a(n) _____ scale.
   a. nominal
   b. ordinal
   c. interval
   d. ratio

9. A Pearson r calculated from a sample could be due to chance, so it is always necessary to
   a. test it for significance.
   b. plot a scatter diagram.
   c. use the computational formula.
   d. base your correlation on a large sample.

10. When using the Spearman rank-difference method, if you find that the sum of the squared differences is very small, you know that the correlation coefficient is likely to be
    a. low positive.
    b. low negative.
    c. high positive.
    d. high negative.

11. Restriction of range results in
    a. a correlation coefficient that is too high.
    b. a correlation coefficient that is too low.
    c. no change in the coefficient.
    d. underfed beef cattle.

12. If the correlation between musical aptitude and creativity were high positive, we would predict that if Tammy is low in musical aptitude, she would probably score
    a. high on creativity.
    b. average on creativity.
    c. low on creativity.
    d. cannot tell from the above information.

13. If there is a high negative correlation between X and Y, individuals who score below the mean on X will probably score
    a. below the mean of Y.
    b. at the mean of Y.
    c. above the mean of Y.
    d. Cannot tell from the above information.

14. The size of a correlation coefficient tells us about
    a. the amount of the relationship.
    b. the direction of the relationship.
    c. the shape of the scatter diagram.
    d. two of the above.
    e. all of the above.

15. A scatter diagram of a correlation near zero would show
    a. a circular pattern of points.
    b. a narrow ellipse pattern.
    c. points on a horizontal line.
    d. none of the above.

16. You are calculating the Pearson r and find that $\Sigma Z_X Z_Y$ is 9.74.
    You would conclude that the relationship between the two variables is
    a. low.   c. moderate.
    b. high.   d. cannot say.

17. A high negative correlation will occur if large negative z scores in X are paired with
    _____ z scores in Y.
    a. large negative   c. large positive
    b. small negative   d. small positive

18. When there is a correlation of zero, large positive z scores in X are paired with
    _____ z scores in Y.
    a. small negative   c. large positive
    b. small positive   d. large negative
                   e. all of the above

19. The Pearson r may be calculated using the computational formula without knowledge of the
    a. means of X and Y.
    b. standard deviations of X and Y.
    c. number of X, Y pairs.
    d. two of the above.
    e. all of the above.

20. When calculating the standard error of the Pearson r, you note that the size of the standard error depends on
    a. the size of the sample.   c. the sample standard deviation.
    b. the size of the population.   d. two of the above.
                   e. none of the above

21. The Spearman rank-difference method is intended for data that
    a. are at least of the ordinal type.   c. contain less than 30 pairs.
    b. can be ranked.   d. All of the above.

22. If there is a negative correlation between two variables
    a. as one decreases the other increases.
    b. as one decreases the other decreases.
    c. as one increases the other increases.
    d. two of the above are correct.

23. A scatter diagram showing a perfect positive relationship would have
    a. a circular pattern of points.
    b. a narrow ellipse going from the upper left corner to the lower right.
    c. points on a straight line from lower left to upper right.
    d. none of the above.

24. The computational formula for the Pearson r is preferred to the z score method, because it is not necessary to calculate
    a. means.
    b. sum of XY.
    c. standard deviations.
    d. two of the above.
    e. all of the above.

25. The term ΣXY means that
    a. ΣX is multiplied by ΣY.
    b. each X is multiplied by its paired Y, and the products are summed.
    c. each XY is multiplied by itself, and the products are summed.
    d. the X and Y values are summed separately, and these sums are multiplied.

26. To test our Pearson r to see if it might be due to sampling error, we begin by assuming that
    a. the true correlation is 1.0.
    b. the sample r is 1.0.
    c. the true correlation is zero.
    d. the sample is normally distributed.

27. When doing a correlation study, one should always check the data for linearity by using a
    a. scatter diagram.
    b. correlation coefficient.
    c. regression equation.
    d. standard error of estimate.

28. To calculate the Spearman rank-difference coefficient, one must have data that are at least _____ in nature.
    a. nominal
    b. ordinal
    c. interval
    d. ratio

29. When ranking the scores in a column while calculating the Spearman rank-difference method, you notice that the four highest scorers each had a score of 53. In your column of ranks, you would assign each of these four a rank of
    a. 1.
    b. 2.5.
    c. 1.5.
    d. 4.

30. A friend of yours is a school psychologist who finds a high positive correlation between mental age and weight in elementary school children. We would conclude that she has
    a. made a computational error.
    b. demonstrated cause and effect.
    c. found a correlation that is really due to chance.
    d. demonstrated the effect of a third variable.

31. In a sample of college men, a correlation coefficient of .50 between height and weight would tell us that
    a. 50 percent of the sample were both tall and heavy.
    b. there was a significant relationship between height and weight.
    c. as height increased, weight also increased.
    d. Two of the above.
    e. All of the above.

32. In a sample of college men, a correlation coefficient of .50 between height and weight would tell us that
    a. as height increased, weight increased.
    b. as height increased, weight decreased.
    c. as height decreased, weight decreased.
    d. Two of the above.

33. The correlation coefficient showing the <u>least</u> amount of relationship would be
    a. -.30   c. -1.0
    b. .22   d. .10

34. A correlation coefficient of +1.04 would indicate a
    a. high degree of relationship.   c. low degree of relationship.
    b. moderate degree of relationship.   d. computational error.

35. A scatter diagram showing plotted points in an ellipse going from lower left to upper right would indicate _____ correlation between X and Y.
    a. positive   c. zero
    b. negative   d. cannot be determined

36. A scatter diagram showing plotted points in an ellipse going from upper left to lower right would indicate _____ correlation between X and Y.
    a. positive   c. zero
    b. negative   d. cannot be determined

37. Plotted points in a very narrow ellipse going from upper left to lower right would indicate _____ correlation between X and Y.
    a. high positive   c. high negative
    b. low positive   d. low negative

38. Plotted points in a very narrow ellipse going from lower left to upper right would show _____ correlation between X and Y.
    a. high positive
    b. low positive
    c. high negative
    d. low negative

39. If the z score method is used to calculate a Pearson $r = .95$, we would expect
    a. large positive z's in X paired with large positive z's in Y.
    b. large positive z's in X paired with large negative z's in Y.
    c. large negative z's in X paired with large negative z's in Y.
    d. two of the above.
    e. none of the above.

40. After calculating a Pearson r, it is necessary to determine
    a. if it is significantly different from zero.
    b. if it could be due to chance.
    c. if it could be due to sampling error.
    d. all of the above.

41. In using Table B, we find that for a sample size of 25 (df = 23) the tabled value for the .05 level is .396. If we assume H_0 that the population correlation is zero, this means that in the sampling distribution.
    a. approximately 40 percent of the r's are significant.
    b. 95 percent of the sample r's fall between $\pm .396$.
    c. the true correlation is .396.
    d. None of the above.

42. We use Table B instead of calculating the standard error of r to determine the confidence interval because
    a. with small samples the sampling distribution of r is non-normal.
    b. with large samples the sampling distribution of r is non-normal.
    c. the standard error of r cannot always be calculated.
    d. Two of the above.

43. Restricting the range of scores _____ the size of a Pearson r.
    a. increases
    b. decreases
    c. has no effect on
    d. effect is not predictable

44. In testing a Pearson r for significance, you note that the tabled value at the .05 level is .40. If your sample yielded an r of -.43, you would decide to
    a. reject H_0 that the population r is zero.
    b. not reject H_0 that the population r is zero.
    c. reject H_0 that the population r is .43.
    d. not reject H_0 that the population r is .43.

51

45. As a positive Pearson r increases in size, we would expect that of those scoring above the median on X, an increasing number should score
    a. above the median on X
    b. below the median on X
    c. above the median on Y
    d. below the median on Y

CHAPTER 8
PREDICTION AND REGRESSION

1. The equation for a straight line is $Y = bX + a$, where b is called the:
   a. regression factor.
   b. correlation coefficient
   c. slope.
   d. intercept.

2. In the equation $Y = bX + a$, a is the
   a. regression factor.
   b. correlation coefficient.
   c. slope.
   d. intercept.

3. The standard error of estimate is a measure of:
   a. variation in Y.
   b. variation of successive sample r values.
   c. variation of Y values around the regression line.
   d. variation of Y values around the mean.

4. Which is <u>not</u> an assumption involved in using the standard error of estimate?
   a. Linearity of regression
   b. Homoscedasticity
   c. Normality of X and Y populations
   d. Ratio data.

5. The regression equation
   a. can be used to predict Y from a given value of X.
   b. is the equation of the best fitting straight line on a scatter diagram.
   c. is of the form $Y = bX + a$.
   d. Two of the above.
   e. All of the above.

6. The ratio of the variance in Y that is due to differences in X to the total variance in Y is known as the
   a. regression coefficient.
   b. coefficient of determination.
   c. correlation coefficient.
   d. regression variance.

7. The regression equation derives its name from the "law of filial regression," a study of inheritance by
   a. Galton.
   b. Pearson.
   c. Spearman.
   d. Darwin.

8. The "best fitting straight line" is such that the squared differences between _____ and _____ are a minimum.
   a. Y; Y′
   b. Y; $\overline{Y}$
   c. X; Y
   d. Y; Y²

9. The error of estimate, e, is defined as
   a. the difference between an observed and a predicted value of Y.
   b. the difference between an observed value of Y and its corresponding value on the regression line.
   c. Y minus Y′.
   d. Two of the above.
   e. All of the above.

10. When we predict Y from a given value of X, the degree of accuracy of our prediction is directly related to the size of the
    a. correlation coefficient.
    b. standard deviation of X.
    c. standard deviation of Y.
    d. All of the above determine prediction accuracy.

11. If there were a perfect correlation between X and Y, we would note that the standard error of estimate, s_E, would be
    a. relatively large
    b. relatively small
    c. zero
    d. a negative value

12. When r = 0, the regression line is a
    a. horizontal line at the mean of X.
    b. horizontal line at the mean of Y.
    c. diagonal line at 45 degrees.
    d. a curved line at the mean of X.

13. If a correlation between reading ability and visual acuity is .40, we could say that _____ percent of the variation in reading scores is due to differences in students' vision.
    a. 16
    b. 32
    c. 40
    d. 84

14. Homoscedasticity literally means
    a. lack of skewness.
    b. X is equal to Y.
    c. decreasing variability.
    d. equal spread.

15. Homoscedasticity is a characteristic of data if the
    a. plot of X and Y values in a scatter diagram is linear.
    b. observed values of Y are normally distributed about Y'.
    c. standard deviation of X is about the same as the standard deviation of Y.
    d. None of the above.

16. In the equation for the straight line Y = 2X + 3, the slope of the line is such that for an increase of one unit in X there is an increase of _____ in Y.
    a. .67
    b. 2.0
    c. 1.0
    d. 3.0

17. In the equation for the straight line Y = 4X + 3, the line intercepts the Y axis at a Y of
    a. .75.
    b. 3.
    c. 4.
    d. Cannot be determined from the above information.

18. The standard error of estimate, s_E, is a standard deviation of which distribution?
    a. Predicted values of Y.
    b. Observed values of Y for a given value of X.
    c. All observed values of Y.
    d. All values of X.

19. All of the following assumptions are necessary for the Pearson r in prediction, except which one?
    a. sample size greater than 30.
    b. normality of X and Y populations.
    c. homoscedasticity.
    d. linear regression.
    e. All of the above are necessary assumptions.

20. In the equation Y = 2X + 4, the line intercepts the Y axis at a Y of
    a. .5.
    b. 2.
    c. 4.
    d. 8.
    e. Cannot be determined from the above information.

21. In the equation Y = 3X + 2, the slope of the line is such that for an increase of one unit in X there is an increase of _____ in Y.
    a. .67
    b. 2.0
    c. 1.0
    d. 3.0

22. When the standard error of estimate, s_E, is large, we would expect to find
    a. small Y - Y′ differences.
    b. small X - Y differences.
    c. large Y - Y′ differences.
    d. large X - Y differences.

23. When r = 0, the predicted value of Y′ is always
    a. zero.
    b. the same as the given value of X.
    c. the same as the mean of X
    d. the same as the mean of Y.

CHAPTER 9
THE SIGNIFICANCE OF THE DIFFERENCE BETWEEN MEANS

1. The mean of a distribution of differences between pairs of sample means drawn from the same population would be expected to have a value
   a. of one.
   b. of zero.
   c. of approximately 30.
   d. Cannot be determined, as we do not know the value of the whole scores.

2. A standard error statistic is basically:
   a. a measure of central tendency.
   b. a variance.
   c. a measure of location.
   d. a standard deviation.

3. When testing the significance of the difference between two means, a value is calculated which indicates:
   a. the probability of a Type I error.
   b. the probability of rejecting the null hypothesis.
   c. the probability of the obtained difference occurring by chance.
   d. Both a and c are correct.

4. Which of the following statements is true regarding the null hypothesis?
   a. Rejecting the null hypothesis is the same as saying there is no significant difference between sample means.
   b. The null hypothesis asserts that the mean of the sampling distribution of differences is zero.
   c. The null hypothesis asserts that the difference between the two population means is zero.
   d. Both b and c are correct.

5. When samples are drawn from two different populations with different means, but the obtained difference between means is not significant, which of the following has occurred?
   a. A Type I error
   b. A Type II error
   c. H_0 has been rejected
   d. Both a and c

6. To fail to reject the null hypothesis when it is false is:
   a. a Type I error.
   b. a Type II error.
   c. a Type III error.
   d. the power of a test.

7. The probability of a Type I error
   a. is called the power of a statistic.
   b. occurs under conditions of a false null hypothesis.
   c. will occur under conditions where t is small (i.e., near 1).
   d. is under the direct control of the experimenter.

8. If an experimenter made a Type I error, he or she
   a. doesn't reject the null hypothesis when it is true.
   b. doesn't reject the null hypothesis when it is false.
   c. rejects the null hypothesis when it is true.
   d. rejects the null hypothesis when it is false.

9. The power of a statistic is the:
   a. ability of a statistic to reject the null hypothesis when it is false.
   b. ability of a statistic to accept the null hypothesis when it is true.
   c. the degree to which a statistic fulfills the normality of populations assumption.
   d. the degree to which a statistic fulfills the equal variances assumption.

10. The relationship between a Type I and a Type II error is:
    a. as one increases, the other decreases.
    b. virtually nil.
    c. as one increases, the other increases.
    d. as one decreases, the other decreases.

11. The t statistic was developed mainly to deal with:
    a. large samples.   c. small samples.
    b. moderate samples.   d. ordinal data.

12. The use of the t ratio assumes that our data is at least:
    a. nominal.   c. interval.
    b. ordinal.   d. ratio.

13. Degrees of freedom for computing t can be expressed as:
    a. $N_1 + N_2 - 2$ c. $N_1 - N_2 - 2$
    b. $\sqrt{N_1 + N_2 - 2}$ d. $N_1 - N_2 - 1$

14. What are the degrees of freedom for a t-test involving correlated samples?
    a. $(N_1 - 1) + (N_2 - 1)$
    b. $N - 1$, where N is the total number of observations
    c. $N - 1$, where N is the number of pairs of observations
    d. $N_1 + N_2 - 1$

15. Which condition must be met to assert that a one-tailed test has more power than a two-tailed test?
    a. there is a substantial correlation between the pretest and criterion measure.
    b. a larger N is used in a one-tailed test.
    c. the researcher is willing to accept nonsignificant differences.
    d. the direction of the difference is specified in advance.

16. The sampling distribution of differences between means has a mean and standard deviation, respectively, of
    a. μ_{diff} and σ_{diff}.
    b. $\overline{X}_1 - \overline{X}_2$ and s.
    c. μ and σ.
    d. μ_{diff} and σ.

17. A null hypothesis states that $\mu_1 - \mu_2$
    a. is equal to zero.
    b. may take any given value.
    c. is the mean of a sampling distribution.
    d. is a one-tailed test.

18. A z score of 2.13
    a. will happen by chance less than 5% of the time.
    b. will happen by chance more than 5% of the time.
    c. would not be considered a "rare occurrence."
    d. is definitely due to chance effects.

19. In testing to see if there was a significant difference between means, the following results were obtained: $\overline{X}_1 = 72$, $\overline{X}_2 = 62$, $\sigma_{diff} = 10$.
    After calculating our z score, we would conclude there was
    a. no significant difference, $p > .05$.
    b. a significant difference, $p < .05$.
    c. a significant difference, $p < .01$.
    d. Cannot be determined from the above information.

20. The normal curve and the t distribution become very similar as the size of the
    a. population increases.
    b. sample increases.
    c. population decreases.
    d. sample decreases.

21. In most t distributions, what percent of the area lies to the right of 1.96?
    a. Less than 2.5%
    b. 2.5%
    c. More than 2.5%
    d. Cannot be determined from the above information.

22. The sum of eight numbers must equal 17. How many degrees of freedom are there in choosing the eight numbers?
    a. one
    b. seven
    c. six
    d. sixteen

23. A school psychologist conducting a study on learning disabilities has 12 sixth-graders in an experimental group and each one is paired with a younger sister or brother in a control group. If a correlated t test is used, the number of degrees of freedom would be
    a. 11
    b. 22
    c. 12
    d. 23

24. Which of the following is not an assumption for the t test?
    a. random samples.
    b. interval or ratio data.
    c. normally distributed samples.
    d. homogeneity of variance.
    e. All of the above are necessary assumptions.

25. The symbol, $\overline{D}$, in a correlated t test is
    a. always equal to zero.
    b. the variability of the difference distribution.
    c. an estimate of the variability of the sampling distribution of differences.
    d. the difference between the means of the two groups.

26. A Type I error is said to occur when the researcher
    a. rejects the null hypothesis when it is really false.
    b. rejects the null hypothesis when it is really true.
    c. accepts the null hypothesis when it is really false.
    d. accepts the null hypothesis when it is really true.

27. The power of a test is
    a. its ability to accept the null hypothesis when it should be accepted.
    b. the probability given by alpha.
    c. its ability to reject the null hypothesis when it should be rejected.
    d. the probability given by one minus alpha.

28. You are reading a research report where a t of 1.73 is called significant at the .05 level. Assuming that the calculations are correct, you would know that the researcher used
    a. a one-tailed test.
    b. either a one-or-two tailed test.
    c. a two-tailed test.
    d. Cannot tell from the above information.

29. "In actual practice, it is a rare research situation where we would be willing to relinquish the right to report a significant result opposite to our expectations." Such a statement refers to
    a. null vs. alternate hypotheses
    b. one or two-tailed tests
    c. Type I or Type II errors
    d. statistical vs. practical significance

30. In calculating the confidence interval for the population mean when the population standard deviation is not known, the standard deviation of the sampling distribution is estimated from
   a. $\overline{X}$ and s.
   b. s and N.
   c. $\overline{X}$ and N.
   d. $\mu_{\overline{x}}$ and s.

31. If we assume a null hypothesis, $\mu_1 - \mu_2 = 0$, then we expect that the mean of the sampling distribution of differences between means would be
   a. plus or minus 1.96 or greater.
   b. the same as μ_1.
   c. zero.
   d. none of these.

32. A z score of -1.27
   a. would be considered a "rare occurrence."
   b. will happen by chance less than 5% of the time.
   c. will happen by chance more than 5% of the time.
   d. is definitely due to chance effects.

33. In testing for significant differences, the term σ_{diff} is rarely used because
   a. it is a biased estimate.
   b. t scores are more accurate.
   c. we usually do not know the population mean.
   d. we usually do not know the population standard deviation.

34. σ_{diff} is to s_{diff} as
   a. independent sample is to correlated sample.
   b. unbiased is to biased.
   c. significant is to nonsignificant.
   d. z value is to t value.

35. $(\overline{X}_1 - \overline{X}_2) - 0$ divided by s_{diff} is used to
   a. calculate the mean of the sampling distribution of differences.
   b. locate where in the sampling distribution the difference between means falls.
   c. estimate the standard deviation of the population.
   d. estimate the standard deviation of the sampling distribution.

36. You calculate a t of 2.38 and note that the tabled value for .01 is 3.22 and for .05 it is 2.19. You would conclude that the null hypothesis can be
   a. accepted at the .05 level.
   b. rejected at the .01 level.
   c. rejected at the .05 level.
   d. none of the above.

37. The .05 and .01 significance levels have been chosen because they
    a. are the precise limits of a "rare occurrence."
    b. were dictated by mathematical theory.
    c. represent percentiles in the normal curve.
    d. none of the above – they were arbitrarily chosen.

38. A researcher is studying political conservatism among 8 engineering students and 11 humanities students. The number of degrees of freedom for a t test would be
    a. 17
    b. 19
    c. 18
    d. Cannot be determined from the above information.

39. A correlated t test should be used instead of a t test for independent samples
    a. if each subject is measured twice.
    b. when each subject's identical twin is in a control group.
    c. whenever $N_1 = N_2$.
    d. Two of the above.
    e. All of the above.

40. A developmental psychologist is comparing the moral development of four-year-olds and six-year-olds, and finds $\overline{X}_1 = 32$ and $s_1 = 5$ for one group and $\overline{X}_2 = 27$ and $s_2 = 12$ for the other. If a t test is calculated, which assumption is not being met?
    a. random samples
    b. homogeneity of variance
    c. interval or ratio data
    d. cannot be determined from the above information.

41. If there is a real difference between two variables in the population, the researcher is most concerned about
    a. power.
    b. Type I errors.
    c. one-tailed tests.
    d. correlated data.

42. A Type II error is said to occur when the researcher
    a. rejects the null hypothesis when it is really false.
    b. rejects the null hypothesis when it is really true.
    c. accepts the null hypothesis when it is really false.
    d. accepts the null hypothesis when it is really true.

43. Alpha, the probability of making a Type I error
    a. is determined by chance.
    b. is equal to or less than .05.
    c. depends on the researcher's data.
    d. is under the direct control of the researcher.

44. Regarding one- and two-tailed tests, the author recommends
    a. one-tailed tests whenever possible because of their power.
    b. either one- or two-tailed tests
    c. two-tailed tests except in rare circumstances.
    d. two-tailed tests always.

45. In calculating the confidence interval for the population mean when the population standard deviation is not known, the _____ distribution is used.
    a. t
    b. normal
    c. z
    d. both a and c

CHAPTER 10
DECISION MAKING, POWER, AND EFFECT SIZE

1. A Type I error occurs when
   a. H_0 is true and the researcher does not reject it.
   b. H_0 is true and the researcher rejects it.
   c. H_0 is false and the researcher rejects it.
   d. H_0 is false and the researcher does not reject it.

2. A Type II error occurs when
   a. H_0 is true and the researcher does not reject it.
   b. H_0 is true and the researcher rejects it.
   c. H_0 is false and the researcher rejects it.
   d. H_0 is false and the researcher does not reject it.

3. When we talk about rejecting the null hypothesis when it should be rejected, we are discussing the _____ of a statistic.
   a. effect
   b. power
   c. rejection
   d. testability

4. If H_0 is true, the probability of a Type II error
   a. depends on the probability of a Type I error.
   b. is set by the experimenter.
   c. is greater than the probability of a Type I error.
   d. None of the above.

5. If H_0 is false, the probability of a Type II error depends on
   a. the alpha level set by the researcher.
   b. the distance between the real μ and the hypothesized μ.
   c. the size of the sample used by the researcher.
   d. two of the above.
   e. all of the above.

6. The alpha level chosen by the experimenter determines
   a. the region of rejection for the null hypothesis.
   b. the region of non-rejection for the null hypothesis.
   c. the probability of making a Type II error.
   d. two of the above.
   e. all of the above.

7. As the difference between the hypothesized μ and the real μ increases, the probability of a
   a. Type II error decreases.
   b. Type II error increases.
   c. Type I error decreases.
   d. Two of the above.
   e. All of the above.

8. In a study to assess mathematics comprehension using the Mathematics Comprehension Exam (MCE), Researcher A finds $\overline{X} = 74$, s = 17, N = 153. Researcher B finds $\overline{X} = 76$, s = 17, N = 218. If the real $\mu = 75$, you would conclude that
   a. Researcher A has more power.
   b. Researcher B has more power.
   c. a Type II error is less likely for Researcher A.
   d. A Type II error is more likely for Researcher B.
   e. Two of the above.

9. In another study using the MCE, Researcher C finds $\overline{X} = 76$, s = 18, N = 100, while Researcher D finds $\overline{X} = 74$, s = 21, N = 100. If the real $\mu = 75$, you would conclude that
   a. Researcher C has more power.
   b. Researcher D has more power.
   c. A Type II error is less likely for Researcher C.
   d. A Type II error is more likely for Researcher D.
   e. Two of the above.

10. In exercises 8 and 9, the research situations described show that power and Type II errors are affected by
   a. the size of the standard error of the means.
   b. the size of the real μ.
   c. the difference between H_0 and H_A.
   d. two of the above.
   e. all of the above.

11. Effect size is defined as the difference in standard deviation units between the
   a. sample mean and the population mean.
   b. sample standard deviation and the population standard deviation.
   c. hypothesized mean and the real mean.
   d. sample variance and population variance.
   e. None of the above.

12. The power of a statistic is dependent upon
    a. the alpha level set by the researcher.
    b. the distance between the real μ and the hypothesized μ.
    c. the size of the sample used by the researcher.
    d. two of the above.
    e. all of the above.

13. The expression $(\mu_{real} - \mu_{hyp})/\sigma$ is denoted by the symbol
    a. d.
    b. H_0
    c. t
    d. z.

14. The formula expressed in exercise 13 is used in the _____ sample case.
    a. one
    b. two
    c. three
    d. four

15. An effect size is often estimated
    a. by using the results of previous research.
    b. by determining beforehand what constitutes an important difference.
    c. by using the population variance.
    d. Two of the above.
    e. All of the above.

16. Some conventional "benchmark" effect sizes for researchers to use was proposed by
    a. Cohen.
    b. Fisher.
    c. Spearman.
    d. Tukey.

17. The effect size necessary to be considered "large" by the statistician named in exercise 16 was
    a. .50
    b. .80
    c. 1.0
    d. 1.96

CHAPTER 11
ONE-WAY ANALYSIS OF VARIANCE

1. If an F value of 4.50 is required for significance at the 5 percent level, using the same degrees of freedom, an F to be significant at the 1 percent level:
   a. must be smaller than 4.50.
   c. must be larger than 4.50.
   b. must be less than 1.
   d. would require a larger SS_{wg}.

2. If the null hypothesis is false, one would expect the F ratio to be:
   a. about one.
   b. greater than one.
   c. less than one.
   d. Would depend on degrees of freedom associated with the denominator term of F ratio.

3. In the use of the analysis of variance technique, the total variance in a study is divided into _____ major components.
   a. four
   c. two
   b. one
   d. three

4. The variability of the scores from $\overline{X}_T$ is called:
   a. between groups variability.
   c. variability within groups.
   b. total variability.
   d. MS_{bg}.

5. The variability of the scores from their group $\overline{X}$s is called.
   a. variability within groups.
   c. total variability.
   b. variability between groups.
   d. None of these.

6. If the means of all groups are equal to each other and equal to the total mean:
   a. the sum of squares between groups would equal the sum of squares within.
   b. the sum of squares within groups would be very small compared to the sum of squares for between groups.
   c. there will be no sum of squares for between groups.
   d. the sum of squares between groups will equal the sum of squares for total.

7. The sum of squares between groups is based on:
   a. the deviation of each score about the total mean.
   b. the deviation of each score about its group mean.
   c. the deviation of each group mean about the total mean.
   d. the deviation of the group means about each individual score.

8. The division of a sum of squares by its df yields:
   a. F.
   b. a mean square.
   c. a variance estimate.
   d. both (b) and (c)

9. The df for SS_{wg} is given by:
   a. the number of scores in the experiment minus one.
   b. the number of scores in the experiment minus the number of sample means.
   c. the number of samples in the experiment minus one.
   d. the number of samples in the experiment minus the mean of the total.

10. Twenty-five subjects took part in an experiment. Three groups were set up consisting of 8, 8, and 9 subjects each. An F test was performed. The number of degrees of freedom associated with the numerator and denominator, <u>respectively</u>, were:
    a. 2, 22.
    b. 2, 24.
    c. 22, 24.
    d. 7, 8.

11. The F ratio is actually:
    a. the ratio of two standard deviations.
    b. the ratio of two estimates of the population variance.
    c. the ratio of two SS.
    d. the ratio of total variability to partial variability.

12. If one obtains a statistically significant F ratio, which of the following conclusion(s) can be made?
    a. Each mean is significantly different from each of the other means.
    b. Each mean comes from a different population.
    c. There are significant differences among the means.
    d. Both (b) and (c).

13. The Tukey procedure is often used after an analysis of variance to
    a. find which differences between means are significant.
    b. provide an accuracy check on the calculation of F.
    c. determine whether F is due to sampling error.
    d. test for homogeneity of variance.

14. In calculating the studentized range statistic, q, for equal Ns, you would begin with the
    a. mean differences that are zero.
    b. smallest mean difference.
    c. largest mean difference.
    d. means that are significantly different.

15. Whenever a sum of squares is divided by the appropriate degrees of freedom, we obtain
    a. an unbiased estimate of the population variance.
    b. a biased estimate of the population variance.
    c. the standard deviation of the sample.
    d. an F ratio.

16. In a distribution, a single score, X_1, is part of the <u>total</u> variability, as denoted by
    a. $\overline{X}_1 - \overline{X}_T$.
    b. $X_1 - \overline{X}_1$.
    c. $X_1 - X_2$.
    d. $X_1 - \overline{X}_T$.

17. Part of the variability <u>between</u> groups is given by
    a. $\overline{X}_1 - \overline{X}_T$.
    b. $\overline{X}_1 - \overline{X}_2$.
    c. $X_1 - \overline{X}_T$.
    d. $X_1 - \overline{X}_1$.

18. The ANOVA is such that there will always be
    a. variability between groups.
    b. a larger variability between groups than within groups.
    c. variability between groups and variability within groups.
    d. variability within groups.

19. The term $\sum\limits^{k}\sum\limits^{N_G} (X - \overline{X}_T)^2$ is known as the

    a. total sum of squares.
    b. between groups sum of squares
    c. within group sum of squares.
    d. variance estimate.

20. The term $\sum\limits^{k} N_G(\overline{X}_G - \overline{X}_T)^2$ represents the

    a. total sum of squares.
    b. between groups sum of squares.
    c. within group sum of squares.
    d. variance estimate.

21. The degrees of freedom for the total sum of squares is
    a. $k - 1$.
    b. $N_T - k$.
    c. $N_T - 1$.
    d. $N_G - 1$.

22. The degrees of freedom for the between groups sum of squares is
    a. $k - 1$.
    b. $N_T - k$.
    c. $N_T - 1$.
    d. $N_G - 1$.

23. Under the null hypothesis both MS_{BG} and MS_{WG} are estimates of the
    a. population mean.
    b. population variance.
    c. sample mean.
    d. sample variance.

24. The F test in the ANOVA is a ratio of two
    a. degrees of freedom.
    b. sums of squares.
    c. means.
    d. variance estimates.

25. A business consultant is comparing the keyboard skills of 10 data entry operators, 5 bookkeepers, and 12 clerical assistants. In an ANOVA design, the degrees of freedom for the F test would be
    a. 2, 24
    b. 3, 24
    c. 2, 26
    d. 3, 26

26. Which of the terms below is not shown in the summary table for the analysis of variance?
    a. sums of squares
    b. mean squares
    c. degrees of freedom
    d. sample means
    e. all of the above are included in a summary table.

27. The usual t test should not be used in which of the following?
    a. comparing heart rates of males and females.
    b. comparing Type A behaviors of dentists, engineers and lawyers.
    c. comparing eye-hand coordination of two-year-olds and three-year-olds.
    d. two of the above.
    e. all of the above could be treated with the usual t test.

28. The quantity $\Sigma(X - \overline{X})^2$ cannot be used as a measure of variability because
    a. its value depends on sample size.
    b. it is always equal to zero.
    c. it is a biased estimate.
    d. the degrees of freedom cannot be determined.

29. In an ANOVA, part of the variability within groups is given by
    a. $\overline{X}_1 - \overline{X}_2$.
    b. $\overline{X}_1 - \overline{X}_T$.
    c. $X_1 - \overline{X}_T$.
    d. $X_1 - \overline{X}_1$.

30. The term $\sum\limits^{k}\sum\limits^{N_G} (X - \overline{X}_G)^2$ is called the

    a. total sum of squares.
    b. between groups sum of squares.
    c. within group sum of squares.
    d. variance estimate.

31. The degrees of freedom for the within group sum of squares is
    a. $k - 1$.
    b. $N_T - k$.
    c. $N_T - 1$.
    d. $N_G - 1$.

32. Another term for "variance estimate" in the ANOVA is
    a. sum of squares.
    b. partitioned degrees of freedom.
    c. total variability.
    d. mean square.

33. Under a true null hypothesis, the value of F should be
    a. zero.
    b. one.
    c. greater than one.
    d. Cannot be determined without an F table.

34. In the analysis of variance, the different sums of squares are additive. Which of the following is correct?
    a. $SS_T + SS_{WG} = SS_{BG}$
    b. $SS_{BG} + SS_{WG} = SS_T$
    c. $SS_T + SS_{BG} = SS_W$
    d. $SS_{BG} + SS_{WG} + SS_T = 1.0$

35. Which of the assumptions listed below is not an assumption for the analysis of variance?
    a. homogeneity of variance
    b. interval or ratio data
    c. random samples
    d. normally distributed samples
    e. all of the above are assumptions for the analysis of variance.

CHAPTER 12
TWO-WAY ANALYSIS OF VARIANCE

1. Which of the following requires a two-way ANOVA instead of a one-way ANOVA?
   a. Finding which of three teaching methods is best for a health course.
   b. Determining if men or women have better motor coordination.
   c. Seeing if there is a significant difference in resting heart rate for 6, 8, 12 and 18 year old public school students.
   d. None of the above - - all could be studied by the one-way ANOVA.

2. We prefer to use a two-way ANOVA instead of two separate experiments because
   a. an interaction effect can be studied.
   b. less subjects are needed.
   c. a Type II error is less likely.
   d. three or more groups can be studied.

3. When we use a factorial design, the term "factor" refers to
   a. variables.   c. cells.
   b. levels.   d. samples.

4. In a two by four factorial design there would be _____ cells.
   a. two
   b. four
   c. six
   d. eight
   e. Cannot say from the above information.

5. In a two by four factorial design there would be _____ factors.
   a. two
   b. four
   c. six
   d. eight
   e. Cannot say from the above information.

6. In order to study main effects in a two-way ANOVA, we would compare
   a. cell means.   c. column means.
   b. row means.   d. total means.
                  e. two of the above.

7. When we graph the results of a two-way ANOVA, a <u>nonsignificant</u> interaction is shown when the curves are
   a. parallel.
   b. converging.
   c. diverging.
   d. two of the above.
   e. all of the above.

8. A <u>significant</u> interaction is shown in the graph of the results of a two-way ANOVA when the curves are
   a. parallel.
   b. converging.
   c. diverging.
   d. two of the above.
   e. all of the above.

9. In order to evaluate interaction effects in a two-way ANOVA, we compare
   a. total means.
   b. cell means.
   c. row means.
   d. column means.
   e. two of the above.

10. When the results of a two-way ANOVA are graphed, a separation between two curves may indicate a
    a. significant interaction.
    b. nonsignificant interaction.
    c. significant main effect.
    d. nonsignificant main effect.

11. In the two-way ANOVA $SS_R + SS_C + SS_{RxC}$ is equal to
    a. SS_{WG}.
    b. SS_T.
    c. SS_{BG}.
    d. None of the above.

12. The degrees of freedom for SS_{WG} in the two-way ANOVA is $N_T - rc$, where rc is the number of
    a. factors.
    b. variables.
    c. cells.
    d. subjects.
    e. two of the above.

13. The term $N_T - 1$ is always the degrees of freedom for the
    a. between groups sum of squares.
    b. within groups sum of squares.
    c. total sum of squares.
    d. interaction sum of squares.
    e. None of the above.

14. In a two-way ANOVA the denominator for all the F tests is MS_{WG}, because it is the only variance estimate not affected by
    a. the column variable.
    b. the row variable.
    c. the interaction of the row and column variables.
    d. all of the above.

15. If there is a significant interaction in a two-way ANOVA, we would expect that MS_{RxC} would be _____ MS_{WG}.
    a. larger than
    b. equal to
    c. smaller than
    d. Cannot tell without knowing the degrees of freedom.

16. If the null hypothesis is true in a two-way ANOVA, we would expect MS_C to be _____ MS_{WG}.
    a. larger than
    b. equal to
    c. less than
    d. either b or c, depending on sample size.

17. Which of the following is not an assumption for the two-way ANOVA described in Chapter 12?
    a. Populations from which the samples were drawn have nearly equal variances.
    b. Populations from which the sample came are normal.
    c. Interval or ratio data.
    d. Equal number of observations in each cell.
    e. All of the above are assumptions for the two-way ANOVA described in Chapter 12.

18. Children diagnosed with attention deficit hyperactivity disorder (ADHD) and a control group of children diagnosed as not having the disorder were rated for "creativity of play" in the presence of many toys or when there were just a few toys present. A rating of one indicates low creativity and ten indicates high creativity. The mean ratings for the four groups were:
    ADHD- -Many toys = 4.9; ADHD—Few toys = 1.8.
    Non-ADHD—Many toys = 5.1; Non-ADHD—Few toys = 8.2
    What would your conclusions be? Indicate more than one choice if needed.
    a. Nonsignificant main effect of quantity of toys.
    b. Significant main effect of quantity of toys.
    c. Nonsignificant main effect of diagnosis.
    d. Nonsignificant interaction.
    e. Significant interaction.

19. Men and women perform a manual dexterity test in a noisy room (stress condition) or a quiet room (no stress condition). The mean error scores obtained are:
Men - Stress = 24; Men - No stress = 15;
Women - Stress = 23; Women - No stress = 14.
What might your conclusions be? Indicate more than one answer if needed.
a. Nonsignificant main effect of stress.
b. Significant main effect of stress.
c. Nonsignificant main effect of sex.
d. Nonsignificant interaction.
e. Significant interaction.

CHAPTER 13
SOME NONPARAMETRIC STATISTICAL TESTS

1. Nonparametric statistics find their greatest use when
   a. greater power is needed.
   b. normality and homogeneity assumptions cannot be met.
   c. data are of the nominal or ordinal type.
   d. Two of the above.
   e. All of the above.

2. Twenty cards are drawn (with replacement) from a deck that is shuffled each time, yielding 7 spades, 3 hearts, 4 diamonds, and 6 clubs. A chi square test to see if this distribution is different from chance would have _____ degrees of freedom.
   a. one
   b. two
   c. three
   d. four

3. The chi square tells us the probability of obtaining a chance difference between
   a. two sample means.
   b. two sample medians.
   c. observed and expected frequencies.
   d. observations in two categories (e.g., Republicans and Democrats).

4. There are many chi square distributions, with the shape of each dependent on
   a. the significance level chosen by the researcher.
   b. the degrees of freedom.
   c. the size of the sample.
   d. the size of the 0 minus E differences.

5. For a given chi square distribution, the tabled value at the .01 level is 11.34. This indicates that
   a. 1% of the distribution lies to the right of 11.34.
   b. 99% of the distribution lies to the right of 11.34.
   c. 0.5% of the distribution lies beyond 11.34.
   d. None of the above.

6. A friend of yours is overjoyed to calculate a chi square of 27.89 on some research data. The first thing you want to know before joining the celebration is
   a. the value of the medians.
   b. the size of the expected frequencies.
   c. the row and column totals.
   d. the degrees of freedom.

7. After you have analyzed your research data, you obtain a chi square of 7.83. To see if it is significant, you find a chi square table that lists the .05 value as 9.49. You now
   a. accept the null hypothesis.
   b. reject the null hypothesis at the .05 level.
   c. reject the null hypothesis at beyond the .05 level.
   d. Not enough information to make a decision.

8.

30	40
10	20

In a chi square analysis of the above frequencies, the expected frequency for the upper right cell would be
   a. 28.   c. 42.
   b. 40.   d. 70.

9. A sample of upper-class students (sophomores, juniors and seniors) are classified as to residence (on-campus, apartment, at home, or "other"). For a chi square analysis of this data, how many degrees of freedom are there?
   a. two   c. seven
   b. six   d. twelve

10. Data analysis using the Mann-Whitney results in the _____ statistic.
   a. z   c. chi square.
   b. T   d. U

11. The null hypothesis for the Mann-Whitney test states that
   a. the number of expected frequencies is equal to the number of observed frequencies.
   b. two populations have the same distribution.
   c. p = .50 that a score from one population would be larger than a score from the other population.
   d. Two of the above.

12. The distribution used by the Kruskal-Wallis test is the _____ distribution.
   a. z   c. chi square
   b. T   d. H

13. The Kruskal-Wallis test is unique among tests for independent samples in that
   a. differences among three or more samples can be evaluated.
   b. the data from the samples are ranked.
   c. it is designed for ordinal data.
   d. it evaluates differences between observed and expected frequencies.

14. The null hypothesis for the Kruskal-Wallis test states that
    a. the means of the samples should be about the same.
    b. the distributions of the samples should be similar.
    c. the sums of the combined ranks for each sample should be approximately the same.
    d. the sums of the scores in each sample should be approximately the same.

15. A nonparametric test that is similar to the usual one-way analysis of variance for independent groups is the _____ test.
    a. Wilcoxon.
    b. Kruskal-Wallis.
    c. Mann-Whitney.
    d. Friedman.

16. Forty personnel directors are rated on their awareness of sexism in hiring practices before and again one month after attending a seminar on discrimination during employment interviews. To see if there was a significant decrease in sexist attitudes, you would use the
    a. Mann-Whitney test.
    b. Kruskal-Wallis test.
    c. sign test.
    d. Friedman test.

17. A nonparametric test that is more powerful than the sign test, and can be used when the magnitude of the differences can be measured is the _____ test.
    a. Friedman
    b. Kruskal-Wallis
    c. Mann-Whitney
    d. Wilcoxon

18. Twenty university students majoring in English are given a vocabulary test, a spatial relations test, and a digit-span memory test. To test your hypothesis that these students would score higher on the vocabulary test, you would use the _____ test.
    a. Friedman
    b. Kruskal-Wallis
    c. Mann-Whitney
    d. Wilcoxon

19. The main weakness of nonparametric tests when compared to parametric tests is that
    a. the researcher is more likely to make a Type II error.
    b. more subjects are needed.
    c. the nonparametric tests are less powerful.
    d. All of the above.

20. When we have a choice we should use a parametric statistic rather than a nonparametric. This is because
    a. there are no assumptions necessary for the nonparametric.
    b. the nonparametric can be used only with nominal or ordinal data.
    c. the parametric has more power.
    d. the parametric is based on sampling theory.
    e. All of the above.

ANSWER KEY

CHAPTER 1

1.	D	14.	A	27.	C	40.	D	53.	A
2.	B	15.	D	28.	B	41.	C	54.	B
3.	B	16.	C	29.	D	42.	B	55.	D
4.	D	17.	A	30.	D	43.	A	56.	A
5.	B	18.	A	31.	B	44.	D	57.	C
6.	D	19.	D	32.	C	45.	B	58.	B
7.	D	20.	C	33.	C	46.	A	59.	C
8.	B	21.	A	34.	D	47.	D	60.	C
9.	B	22.	A	35.	D	48.	B	61.	A
10.	C	23.	A	36.	A	49.	B	62.	D
11.	B	24.	A	37.	A	50.	B	63.	C
12.	C	25.	C	38.	A	51.	D	64.	D
13.	C	26.	B	39.	C	52.	B		

CHAPTER 2

1.	A	14.	B	27.	B	40.	D	53.	B
2.	C	15.	D	28.	B	41.	D	54.	B
3.	A	16.	D	29.	C	42.	B	55.	C
4.	A	17.	A	30.	A	43.	A	56.	D
5.	D	18.	C	31.	B	44.	D	57.	C
6.	A	19.	B	32.	A	45.	C	58.	D
7.	D	20.	C	33.	B	46.	A	59.	B
8.	D	21.	C	34.	D	47.	D	60.	A
9.	B	22.	D	35.	A	48.	C	61.	C
10.	C	23.	D	36.	C	49.	B	62.	D
11.	A	24.	D	37.	C	50.	A	63.	C
12.	B	25.	A	38.	A	51.	C	64.	A
13.	C	26.	A	39.	B	52.	C		

CHAPTER 3

1.	B	10.	C	19.	B	28.	C	37.	A
2.	A	11.	B	20.	D	29.	A	38.	D
3.	C	12.	B	21.	A	30.	B	39.	C
4.	B	13.	B	22.	A	31.	C	40.	A
5.	C	14.	D	23.	B	32.	C	41.	C
6.	C	15.	B	24.	C	33.	C	42.	D
7.	B	16.	B	25.	A	34.	B	43.	B
8.	B	17.	C	26.	C	35.	B	44.	D
9.	A	18.	B	27.	A	36.	A	45.	C

CHAPTER 4

1.	B	11.	A	21.	B	31.	A	41.	B
2.	A	12.	C	22.	D	32.	D	42.	B
3.	B	13.	B	23.	B	33.	B	43.	C
4.	B	14.	A	24.	D	34.	C	44.	A
5.	C	15.	D	25.	C	35.	C	45.	D
6.	D	16.	C	26.	C	36.	A	46.	D
7.	C	17.	B	27.	C	37.	D	47.	D
8.	C	18.	A	28.	A	38.	C	48.	B
9.	C	19.	C	29.	B	39.	C	49.	A
10.	C	20.	A	30.	A	40.	D		

50. 68, 132
51. -.75
52. 128
53. 6, 30
54. -.75
55. 23

CHAPTER 5

1.	C	11.	D	21.	B	31.	D	41.	C
2.	E	12.	B	22.	D	32.	C	42.	A
3.	D	13.	B	23.	C	33.	B	43.	A
4.	A	14.	D	24.	D	34.	A	44.	D
5.	B	15.	D	25.	B	35.	C	45.	D
6.	C	16.	D	26.	C	36.	A	46.	D
7.	A	17.	D	27.	A	37.	C	47.	A
8.	A	18.	B	28.	B	38.	D	48.	D
9.	C	19.	A	29.	A	39.	D	49.	D
10.	A	20.	B	30.	B	40.	C	50.	C
								51.	A

52. 6.68%
53. 82.3%
54. 416
55. 304-696
56. .84
57. 628

58. 4.01%
59. 78.88%
60. 125.6
61. 60.8-139.2
62. .69
63. 83.2

CHAPTER 6

1. A	9. D	17. B	25. D	33. D
2. C	10. A	18. B	26. D	34. B
3. C	11. C	19. B	27. C	35. B
4. B	12. B	20. A	28. D	36. B
5. B	13. A	21. D	29. E	37. C
6. C	14. D	22. C	30. B	38. D
7. D	15. C	23. D	31. C	39. A
8. C	16. B	24. A	32. B	

40. 0.3
41. 66.41-67.59
42. No, not in the CI.
43. 0.7
44. 65.63-68.37
45. No, not in the CI.

CHAPTER 7

1. D	10. C	19. D	28. B	37. C
2. C	11. B	20. A	29. B	38. A
3. A	12. C	21. D	30. D	39. D
4. B	13. C	22. A	31. C	40. D
5. D	14. D	23. C	32. D	41. B
6. A	15. A	24. D	33. D	42. A
7. C	16. D	25. B	34. D	43. B
8. C	17. C	26. C	35. A	44. A
9. A	18. E	27. A	36. B	45. C

CHAPTER 8

1. C	6. B	11. C	16. B	21. D
2. D	7. A	12. B	17. B	22. C
3. C	8. A	13. A	18. B	23. D
4. D	9. E	14. D	19. A	
5. D	10. A	15. D	20. C	

CHAPTER 9

1. B	10. A	19. A	28. A	37. D
2. D	11. C	20. B	29. B	38. A
3. D	12. C	21. C	30. B	39. D
4. D	13. A	22. B	31. C	40. B
5. B	14. C	23. A	32. C	41. A
6. B	15. D	24. C	33. D	42. C
7. D	16. A	25. D	34. D	43. D
8. C	17. B	26. B	35. B	44. C
9. A	18. A	27. C	36. C	45. A

CHAPTER 10

1. B	5. E	9. E	13. A	17. B
2. D	6. D	10. A	14. A	
3. B	7. A	11. C	15. D	
4. D	8. B	12. E	16. A	

CHAPTER 11

1. C	8. D	15. A	22. A	29. D
2. B	9. B	16. D	23. B	30. C
3. C	10. A	17. A	24. D	31. B
4. B	11. B	18. D	25. A	32. D
5. A	12. C	19. A	26. D	33. B
6. C	13. A	20. B	27. B	34. B
7. C	14. C	21. C	28. A	35. D

CHAPTER 12

1. D	5. A	9. B	13. C	17. E
2. A	6. E	10. C	14. D	18. A, E
3. A	7. A	11. B	15. A	19. B, C, D
4. D	8. D	12. C	16. B	

CHAPTER 13

1. D	5. A	9. B	13. A	17. D
2. C	6. D	10. D	14. C	18. A
3. C	7. A	11. D	15. B	19. D
4. B	8. C	12. C	16. C	20. C